Au

Ti

AN INTRODUCTION TO POTENTIAL THEORY

UNIVERSITY MATHEMATICAL MONOGRAPHS

Editor
ALAN JEFFREY

1. *Integration, Measure and Probability*	H. R. Pitt
2. *Introduction to Lattice Theory*	D. E. Rutherford
3. *Boundary Value Problems*	A. G. Mackie
4. *Discourse on Fourier Series*	C. Lanczos
5. *The Classical Moment Problem*	N. I. Akhiezer
6. *Lectures on Modern Algebra*	P. Dubreil and M. L. Dubreil-Jacotin
7. *An Introduction to Potential Theory*	Nicolaas du Plessis

AN INTRODUCTION TO POTENTIAL THEORY

NICOLAAS DU PLESSIS

OLIVER & BOYD
EDINBURGH

OLIVER AND BOYD
Tweeddale Court
Edinburgh EH1 1YL

A division of Longman Group Ltd

First published 1970

ISBN 0 05 002054 4

PRINTED IN GREAT BRITAIN BY BELL AND BAIN LTD., GLASGOW

CONTENTS

CHAPTER 4

THE DIRICHLET PROBLEM

PREFACE

Some years ago I had the good fortune to be present at a Colloquium on Potential Theory held at the Orsay division of the University of Paris under the auspices of the Centre Nationale du Recherche Scientifique. The proceedings were opened formally by a member of the C.N.R.S., who began by saying:

"La Théorie du Potentiel est un véritable carrefour de la Mathématique."

This is at once the fascination and the difficulty of the subject. One must perforce use, as tools, results from a number of branches of pure mathematics, and one must also have an appreciation of the applied mathematics which has given rise to the leading ideas in the subject. Kellogg's remarkable book in the Springer series is a *tour de force* in this respect. The book was published in 1929 and Kellogg seems to have imposed on himself a self-denying ordinance in that he decided to keep the pure mathematical requirements for the book as modest as was consistent with both rigour and the ability to give an account of some of the then exciting new developments in the subject. He therefore assumed only knowledge of the Riemann integral, so denying himself the powerful generality of the Lebesgue integral, with which, indeed, he was thoroughly familiar, as his research papers show. Doing so, he succeeded brilliantly in his task since now, nearly forty years later, Kellogg is still required reading. But it was written forty years ago; and a lot can happen, and has happened, in forty years. Indeed, neither Kellogg nor anyone else would have tried to write that particular book much after 1929. One can see this in a rather striking way in an expository article due to Brelot[1]. In 27 pages he devotes a preliminary section of rather less than a page to "La théorie ancienne. Kellogg", and then passes on. So does Time make ancient good uncouth!

Any account of Potential Theory now must ask of the inquirer a great deal more background knowledge. A good theory of integration, effectively incorporating both the Lebesgue and the Lebesgue–Stieltjes integral is essential, if only because one starts by working with semi-continuous functions and these need not be Riemann integrable at all. This and other requirements explain why, in this book, there is a considerable first chapter which sets out the background requirements for the rest of the book; and it is perhaps a mark of the change in the general level of mathematical sophistication over the period, that Kellogg was quite happy to set his theory in Euclidean three-dimensional space, but now one finds it necessary to start by setting up in business in a locally compact Hausdorff space!

About the reason and purpose of Chapter 1 I have already spoken. Chapter 2 sets out the basic theory of classical harmonic and superharmonic functions. Here I have followed, for the most part, an exposition due to Brelot[2] but I have taken the opportunity to combine it with the theory at the Alexandroff point, a happy idea due to Brelot[4] which, among other things, enables one to deal with the interior and the exterior Dirichlet Problem together.

Chapter 3 treats of the Conductor Problem. Here I am happy to acknowledge my debt to Landkov's book, since my account of α-capacity is largely based on the excellent treatment given there. The setting for most of the chapter is $\mathbb{R}^p$, and it should be mentioned that the concepts of energy and capacity have been considerably generalised by Fuglede so that they can be applied to more general kernels defined in locally compact spaces.

The abstract theory in Chapter 4 is based almost wholly on the appropriate sections in Brelot[3] and is peculiarly well-adapted for application to elliptic partial differential equations. In Bauer there will be found a variant version, in which Axiom P is replaced by a separation axiom and which has been developed so as to apply to parabolic differential equations also.

Throughout the book Theorems and Lemmas are numbered in the order in which they appear within each chapter. Thus Theorem 2.18 is the eighteenth result in Chapter 2 and Lemma 3.9 is to be found in Chapter 3. It is never the case that a lemma and a theorem have the same number. Books and articles referred to in the text are identified in the list of references at the end. If only one item by a given author is referred to, reference is only by the name of the author; if there is more than one, I speak of, for example, Brelot[2].

As the title of my book implies, my aim is modest. I seek to introduce the graduate student to Potential Theory. This work does not claim to be definitive in any sense, and it leaves out far more than it puts in. Those versed in the subject will complain, with justice, that there is much main line potential theory which is not represented here at all—notably perhaps that sector of the subject where the connection between potential-theoretical ideas and the theory of stochastic processes is explored by workers such as Doob, Hunt, Mayer, Herz and others. My hope is, nevertheless, that I have made some of the leading ideas in the subject a little more accessible than they have been heretofore.

In conclusion I should like to express my gratitude to Professor Alan Jeffrey, the editor of this series, for encouraging me to write the book, and to Oliver and Boyd for the great pains they have taken to prepare it for printing.

Chapter 1

PRELIMINARIES

In this chapter I gather together some of the prerequisites for later chapters of the book. For the most part, I content myself with the statement of results, referring the reader to other sources for proofs and amplification.

§ 1.1. Semi-continuous functions

Let X be a locally compact Hausdorff space. A function $f : X \to \mathbb{R}^*$ (see footnote) is said to be **lower semi-continuous** (this will be abbreviated to l.s.c.) at $a \in X$, if

$$\liminf_{x \to a} f(x) \geqslant f(a).$$

Similarly, a function $f : X \to \mathbb{R}_*$ is said to be **upper semi-continuous** (abbreviated to u.s.c.) at $a \in X$, if

$$\limsup_{x \to a} f(x) \leqslant f(a).$$

Clearly, f is u.s.c. at a if, and only if, $-f$ is l.s.c. at a. Clearly, also, a function which is both l.s.c. and u.s.c. at a is continuous there.

The **support,** supp f, of a function $f : X \to \overline{\mathbb{R}}$ is defined by

$$\operatorname{supp} f = \text{closure of } \{x \in X \mid f(x) \neq 0\}.$$

If supp f is compact, then f is said to be of **compact support.**

The extended real field, $\overline{\mathbb{R}}$, is the structure obtained by adjoining to the real field $\mathbb{R}$, the ideal numbers ∞ and $-\infty$ and making the operational conventions

$$a+\infty = \infty,\ a-\infty = -\infty \text{ for all } a \in \mathbb{R};$$
$$a\infty = \infty a = \infty, \text{ if } a > 0;\ -\infty \text{ if } a < 0;\ = 0 \text{ if } a = 0;$$
$$\infty+\infty = \infty,\ -\infty-\infty = -\infty; \text{ and } \infty-\infty = 0.$$

Also $\quad -\infty < a < \infty$ for all $a \in \mathbb{R}$, and $\infty^{-1} = 0$.

A function $f : X \to \mathbb{R}$ is said to be *real-valued,* while a function $f : X \to \overline{\mathbb{R}}$ is said to be **numerically-valued.** We denote $\overline{\mathbb{R}} \setminus (-\infty)$ by $\mathbb{R}^*$ and $\overline{\mathbb{R}} \setminus (\infty)$ by $\mathbb{R}_*$.

We now turn to some elementary properties of semi-continuous functions.

THEOREM 1.1. *Suppose that* f, g *are l.s.c. (u.s.c.) in a set* S. *Then* $f+g$ λf with $\lambda \geqslant 0$, *max* $[f, g]$ *and min* $[f, g]$ *are l.s.c. (u.s.c.) in* S.

The proofs, which are quite straightforward, can safely be left to the reader.

THEOREM 1.2. *A function l.s.c. (u.s.c.) in a compact set* K *is bounded below (above) in* K.

For, given $a \in K$, we can find a neighbourhood $N(a)$ such that, for $x \in N(a)$

$$g(x) > g(a)-1.$$

The family $\{N(a)\}_{a \in K}$ is an open covering of K; and so, since K is compact, we can find a finite sub-covering $\{N(a_i)\}_{1 \leqslant i \leqslant k}$. Then, given $x \in K$, we may find i so that $g(x) > g(a_i)-1 \geqslant \min_{1 \leqslant i \leqslant k} g(a_i)-1$.

If f is u.s.c. then $-f$ is l.s.c. and so bounded below. Thus f is bounded above.

THEOREM 1.3. *A necessary and sufficient condition that* f *be l.s.c. (u.s.c.) in* X *is that, for each* $\lambda \in \mathbb{R}$, *the set* $\{x \in X \mid f(x) > \lambda\}$ $[\{x \in X \mid f(x) < \lambda\}]$ *be open.*

Suppose first that f is l.s.c. in X, and let $\lambda \in \mathbb{R}$ and $a \in G = \{x \in X \mid f(x) > \lambda\}$. Then $f(a) > \lambda$, and so there is a neighbourhood N of a such that, for $x \in N$, $f(x) > \lambda$. Thus, given $a \in G$, there is a neighbourhood N such that $a \in N \subset G$. Consequently G is open.

On the other hand, suppose that, for each $\lambda \in \mathbb{R}$, G is open. Let $a \in X$, and let $\varepsilon > 0$ be given. Then $G = \{x \in X \mid f(x) > f(a)-\varepsilon\}$ is open and $a \in G$. So there is a neighbourhood N, such that $a \in N \subset G$. Thus, for $x \in N, f(x) > f(a)-\varepsilon$, and so f is l.s.c. at a.

This proves both the necessity and sufficiency of the condition for lower semi-continuity.

Further, $\{x \in X \mid f(x) < \lambda\} = \{x \in X \mid -f(x) > -\lambda\}$. Now f is u.s.c. in X if, and only if, $-f$ is l.s.c. in X; and this is so, by what we have just proved, if, and only if, the above set is open for every $\lambda \in \mathbb{R}$.

It is both important and fortunate that a locally compact space has a plentiful supply of continuous functions defined on it. This is ensured by Urysohn's Lemma of which one version is

THEOREM 1.4. *Given a compact subset* K *of* X *and an open set* $G \supset K$, *there is a function* $h : X \to \mathbb{R}$ *which is continuous on* X *and such that* $h(K) = 1$, *supp* $h \subset G$ *and* $h(X) \subset [0, 1.]$.

A proof is to be found in Rudin (p. 39, 2.12).

It is also frequently helpful to be able to extend a given continuous function. The following, which is a muted version of Tietze's Extension Theorem, will be of particular value subsequently.

THEOREM 1.5. *Suppose that X is compact and that f is defined and continuous on a compact subset K of X. Then there is an extension f^* of f defined and continuous on X such that $f^*(x) = f(x)$, $x \in K$ and* $\sup_X f^* = \sup_K f$.

A proof is to be found in Rudin (p. 385, 20.4).

Since a continuous function is one which is both l.s.c. and u.s.c., we can, in Theorem 1.1, replace 'l.s.c.' by 'continuous'. But it is well-known that if $(f_\alpha)_{\alpha \in A}$ is a family of continuous functions it is not necessarily the case that either $\sup_A f_\alpha$ or $\inf_A f_\alpha$ need be continuous. In that case $\sup_A f_\alpha$ is l.s.c. and $\inf_A f_\alpha$ is u.s.c. and, indeed, something more is true.

THEOREM 1.6. *Let $\{f_\alpha\}_{\alpha \in A}$ be a family of numerically-valued functions, each of which is l.s.c. (u.s.c.) in X. Then* $\sup_A f_\alpha$ ($\inf_A f_\alpha$) *is l.s.c. (u.s.c.) in X.*

Let $f(x) = \sup_{\alpha \in A} f_\alpha(x)$ and let $a \in X$. Then, assuming $f(a) \in \mathbb{R}$, and given $\varepsilon > 0$, there is $\alpha_0 \in A$ such that

$$f_{\alpha_0}(a) > f(a) - \tfrac{1}{2}\varepsilon.$$

Furthermore, there is a neighbourhood $N(a)$ such that, for $x \in N(a)$,

$$f_{\alpha_0}(x) > f_{\alpha_0}(a) - \tfrac{1}{2}\varepsilon.$$

Since $f(x) \geqslant f_{\alpha_0}(x)$ for all $x \in X$, we have, for $x \in N(a)$,

$$f(x) > f(a) - \varepsilon,$$

so that f is l.s.c. at a and hence in X. A similar proof disposes of the case in which $f(a) = \infty$.

Recalling that f is u.s.c. if, and only if, $-f$ is l.s.c., the u.s.c. half follows easily.

The converse of this last result about families of continuous functions is true in a sense which will be indicated later.

Monotonic sequences of functions have a very important role in classical analysis. In the more abstract context, however, they need to be replaced by up-directed and down-directed families. A family $(f_\alpha)_{\alpha \in A}, f_\alpha : X \to \overline{\mathbb{R}}$ such that, given $\alpha, \beta \in A$ there is a $\gamma \in A$, such that $f_\gamma \geqslant \max[f_\alpha, f_\beta]$ is said to be an *up-directed* family on X. Similarly, a family such that, given $\alpha, \beta \in A$, there is a $\gamma \in A$ such that $f_\gamma \leqslant \min [f_\alpha, f_\beta]$ is said to be *down-directed* family on X.

We may now state

THEOREM 1.7. (1) *A necessary and sufficient condition that a function be l.s.c. on a locally compact space X is that it be the supremum of an up-directed family of functions, each of which is continuous in X.*

(2) *If X is compact then 'supremum of an up-directed family' in (1) may be replaced by 'limit of an increasing sequence'.*

The sufficiency of both (1) and (2) follows from Theorem 1.6.

Turning to the necessity, suppose first that X is compact and that g is l.s.c. on X. Then, by Theorem 1.2, g is bounded below in X. Given $\varepsilon > 0$, let $m = \inf_X [g(x) - \varepsilon]$ and set

$$\mathscr{F} = \{f\colon X \to \mathbb{R} \mid f(x) \leqslant g(x) \text{ and } f \text{ continuous in } X\}.$$

By Urysohn's Lemma, there is a function f continuous in X and such that $f(a) = g(a) - \varepsilon$, $f(X \setminus N(a)) = m$ and $f(X) \subset [m, g(a) - \varepsilon]$. Consequently $f \in \mathscr{F}$; and, since ε is arbitrary, we have, for all $a \in X$,

$$g(a) = \sup [f(a) \mid f \in \mathscr{F}].$$

We now show that we may select from $\mathscr{F}$ a suitable sequence. Given $a \in X$ and any integer n, there is an $f_a^n \in \mathscr{F}$ such that

$$f_a^n(a) > g(a) - 1/2n.$$

Since f_a^n is continuous and g is l.s.c., there is a neighbourhood $N^n(a)$ such that, when $x \in N^n(a)$, $f_a^n(x) > g(x) - 1/n$.

From the open covering $\{N^n(a)\}_{a \in X}$ of X we may select a finite covering $\{N_0(a_r)\}_{1 \leqslant r \leqslant k_n}$. Setting

$$f_n = \sup_{1 \leqslant r \leqslant k_n} f_{a_r}^n,$$

we have that $f_n(x) > g(x) - 1/n$ for all $x \in X$ and $f_n \in \mathscr{F}$.
Finally $g_n = \sup_{1 \leqslant r \leqslant n} f_r \in \mathscr{F}$, and $g(x) - 1/n < g_n(x) \leqslant g(x)$ for $x \in X$.
Furthermore, the sequence $\{g_n\}$ is increasing. This then proves (2).

Returning to (1), suppose that X is locally compact; let $a \in X$ and let K be a compact containing a. Let

$$\mathscr{F}_K = \{f \mid f \text{ is continuous and } f \leqslant g \text{ in } K\},$$
$$\mathscr{F} = \{f \mid f \text{ is continuous, supp} f \text{ is compact and } f \leqslant g \text{ in } X\}.$$

Then $g(a) = \sup_{\mathscr{F}K} f(a) \leqslant \sup_{\mathscr{F}} f(a)$ and $g(a) \geqslant \sup_{\mathscr{F}} f(a)$. Hence $g(a) = \sup_{\mathscr{F}} f(a)$ and, clearly, $\mathscr{F}$ is an up-directed family.

Theorem 1.7 has the counterpart

THEOREM 1.8. (1) *A necessary and sufficient condition that a function be u.s.c. on a locally compact space is that it be the infimum of a down-directed family of functions each of which is continuous in the space X.*

(2) *If X is compact then 'infimum of a down-directed family' in* (1) *may be replaced by 'limit of a decreasing sequence'.*

It is of great value to know that a continuous function can be uniformly approximated by functions from a sub-class of the continuous functions which is significant in a given context. A result of great generality and power in this connection is the Stone Approximation Theorem, of which the following is a variant particularly suited to our purpose.

Let $\mathscr{C}(X)$ denote the class of real-valued functions continuous in X. If X is compact, then we can define a metric d on $\mathscr{C}(X)$ by

$$d(f, g) = \sup_{x \in X} |f(x) - g(x)|,$$

and then $\mathscr{C}(X)$ becomes a metric space. Also, if $f \in \mathscr{C}(X)$ can be uniformly approximated by functions from a sub-class $\mathscr{A}$ of $\mathscr{C}(X)$, this is tantamount to saying that $f \in \overline{\mathscr{A}}$, where the topology in question is the metric topology just defined.

We may define $f+g, fg$ and λf, where $f, g \in \mathscr{C}(X)$ and $\lambda \in \mathbb{R}$ by

$$(f+g)(x) = f(x)+g(x), (fg)(x) = f(x)\, g(x), (\lambda f)(x) = \lambda f(x).$$

Then, clearly, $\mathscr{C}(X)$ is closed under these operations, and any subset of $\mathscr{C}(X)$ which is also so closed is said to be a **sub-algebra** of $\mathscr{C}(X)$.

THEOREM 1.9. *Let X be compact and let $\mathscr{A}$ be a sub-algebra of $\mathscr{C}(X)$ which has the following properties:*

(*a*) *If $f, g \in \mathscr{A}$, then $\sup(f, g), \inf(f, g) \in \mathscr{A}$.*

(*b*) *The function e given by $e(x) = 1, x \in X$ belongs to $\mathscr{A}$.*

(*c*) *$\mathscr{A}$ separates X; that is, given $x, y \in X$, there is $f \in \mathscr{A}$ such that $f(x) \neq f(y)$. Then $\overline{\mathscr{A}} = \mathscr{C}(X)$.*

Let $x, y \in X$ and $a, b \in \mathbb{R}$. Then we can find $g \in \mathscr{A}$ such that $g(x) = a$ and $g(y) = b$. For there is a function $f \in \mathscr{A}$ such that $f(x) \neq f(y)$. Choose $\lambda, \mu \in \mathbb{R}$, as we may, so that

$$\lambda f(x) + \mu = a, \quad \lambda f(y) + \mu = b.$$

Then $g = \lambda f + \mu e \in \mathscr{A}$ and is the required function.

Let $F \in \mathscr{C}(X)$ and let $x, y \in X$. Let $g_{xy} \in \mathscr{A}$ be such that

$$g_{xy}(x) = F(x), \; g_{xy}(y) = F(y).$$

Given $\varepsilon > 0$, there is a neighbourhood $N(y)$ of y such that

$$-g_{xy}(z) > F(z)-\varepsilon \quad \text{for } x \in N(y). \tag{1.1}$$

Hold x fixed, and for each $y \in X$ select an open neighbourhood, $N(y)$, satisfying (1). Then $\{N(y)\}_{y \in X}$ is an open covering of X; and, since X is compact, we may select a finite covering $\{N(y_i)\}_{(i=1,2,\ldots,n)}$ of X.

Let

$$h_x = \sup\{g_{xy_1}, g_{xy_2}, \ldots, g_{xy_n}\}.$$

Then $h_x(x) = F(x)$ and, by (1.1),

$$h_x(z) > F(z)-\varepsilon \text{ when } z \in X. \tag{1.2}$$

Now, by (a), $h_x \in \mathscr{A} \subset \mathscr{C}(X)$, and so we can find a neighbourhood $N'(x)$ of x such that

$$h_x(z) < F(z)+\varepsilon, \text{ when } z \in N'(x).$$

Again $\{N'(x)\}_{(x \in X)}$ is an open covering of X, from which we may select a finite subcovering $\{N'(x_j)\}_{(j=1,2,\ldots,m)}$.

Let $h = \inf(h_{x_1}, \ldots, h_{x_m})$. Then $h \in \mathscr{A}$, and

$$h(z) < F(z)+\varepsilon \text{ when } z \in X. \tag{1.3}$$

Also, from (1.2),

$$h(z) > F(z)-\varepsilon \text{ when } z \in X;$$

and so, for $z \in X$, $|h(z)-F(z)| < \varepsilon$.

Thus, given any ε-neighbourhood

$$B_\varepsilon(F) = \{f \in \mathscr{C}(X) \mid d(f, F) < \varepsilon\}$$

of F in $\mathscr{C}(X)$ we can find $h \in \mathscr{A}$ such that $h \in B_\varepsilon(F)$. Hence $F \in \bar{\mathscr{A}}$, and thus $\bar{\mathscr{A}} = \mathscr{C}(X)$.

When X is a compact subset of $\mathbb{R}^p$, the concept of **polynomial in X** is meaningful, and we may ask whether a continuous function can be uniformly approximated by polynomials. The affirmative answer is the content of the Weierstrass Approximation Theorem.

THEOREM 1.10. *Let $K \subset \mathbb{R}^p$ be compact and let f be continuous in K. Then f can be uniformly approximated by polynomials.*

We remark first that, for $0 \leqslant u \leqslant 1$, the binomial series $\sum_{r=0}^{\infty} \binom{\frac{1}{2}}{r} (-u)^r$ converges uniformly to $(1-u)^{\frac{1}{2}}$. Consequently, given $\varepsilon > 0$, and setting $u = 1-t^2$ we can find n such that

$$\left| \sum_{r=0}^{n} \binom{\frac{1}{2}}{r} (t^2-1)^r - |t| \right| < \varepsilon \text{ for } t \in [-1, 1],$$

and thus, in the interval $[-1, 1]$, $|t|$ can be uniformly approximated by polynomials in t. Since, for $\eta > 0$, $|t| = \eta\,|\frac{t}{\eta}|$ it follows that $|t|$ can be uniformly approximated in $[-\eta, \eta]$ by polynomials in t.

Let p be a polynomial in the variables $x_1, x_2, \ldots x_p$. Then, with $\eta = \sup_K |p(x)|$, $|p(x)|$ can be uniformly approximated in K by polynomials in $p(x)$ and so by polynomials in $x_1, \ldots, x_p$. Thus, if $\mathscr{P}$ is the set of all polynomials in K then, when $p \in \mathscr{P}$, $|p| \in \overline{\mathscr{P}}$.

It is easily verified that $\mathscr{P}$ and $\overline{\mathscr{P}}$ are sub-algebras of $\mathscr{C}(X)$ and that, when $q \in \overline{\mathscr{P}}$, $|q| \in \overline{\mathscr{P}}$. Now

$$\sup(q_1, q_2) = \tfrac{1}{2}(q_1+q_2+|q_1-q_2|)$$
$$\inf(q_1, q_2) = \tfrac{1}{2}(q_1+q_2-|q_1-q_2|),$$

and so, when $q_1, q_2 \in \overline{\mathscr{P}}$, $\sup(q_1, q_2)$ and $\inf(q_1, q_2) \in \overline{\mathscr{P}}$. Furthermore, $e \in \mathscr{P} \subset \overline{\mathscr{P}}$. Finally, if $a, b \in K$, then $|x-a|^2$ is a polynomial taking different values at a and b when $a \neq b$ and so $\mathscr{P}$ (and hence $\overline{\mathscr{P}}$) separates points.

Thus $\overline{\mathscr{P}}$ satisfies the requirements of Theorem 1.9, and so $\overline{\mathscr{P}} = \overline{\overline{\mathscr{P}}} = \mathscr{C}(X)$. This gives Theorem 1.10.

There are, of course, many proofs of Theorem 1.10. Perhaps the most direct one is due to Kellogg (p. 321). It is certainly remarkably free from fuss and any requirement to meddle in the 'higher things'.

§ 1.2. Measure and Integral

Suppose again that X is a locally compact Hausdorff space, and let $\mathfrak{B}$ be the Borel tribe on X, that is, suppose that $\mathfrak{B}$ is the smallest family of subsets of X such that

(*a*) $X \in \mathfrak{B}$.

(*b*) If $B \in \mathfrak{B}$ then $X \setminus B \in \mathfrak{B}$.

(*c*) If $B_i \in \mathfrak{B}$ $(i = 1, 2, \ldots)$, then $\bigcup_{i=1}^{\infty} B_i \in \mathfrak{B}$.

(*d*) If $G \subset X$ is open, then $G \in \mathfrak{B}$.

A mapping $\mu: \mathfrak{B} \to \mathbb{R}^*$ such that

(*i*) $\mu(B) \geqslant 0$ for all $B \in \mathfrak{B}$.

(*ii*) If $B_i \in \mathfrak{B}$, $i = 1, 2 \ldots$ and B_i are disjoint then $\mu(\bigcup_{i=1}^{\infty} B_i) = \sum_{i=1}^{\infty} \mu(B_i)$.

(*iii*) $\mu(B) = \inf\{\mu(G) \mid G \supset B \text{ and } G \text{ open}\}$.

(*iv*) If $K \subset X$ is compact, then $\mu(K) < +\infty$.

is said to be a Radon measure† on X.

† I follow the usage introduced by Landkov. For him a measure is a **non-negative** completely additive set function. A real-valued completely additive set function he terms a **charge.**

The support, supp μ, of a measure μ is defined by

$$\text{supp } \mu = \{\cap(X \setminus G) \mid G \text{ open and } \mu(G) = 0\}.$$

Consequently supp μ is the least closed set F, such that $\mu(F) = \mu(X)$.

A function $f : X \to \overline{\mathbb{R}}$ is said to be measurable or, more precisely, Borel measurable, if for every $\lambda \in \mathbb{R}$ the set $\{x \in X, f(x) > \lambda\} \in \mathfrak{B}$. It follows that, if f and g are measurable, so are the functions $\lambda f + \mu g$, $\lambda, \mu \in \mathbb{R}$ and fg. Also, if f is measurable so is $|f|$. Furthermore, if $\{f_n\}$ is a sequence of functions each of which is measurable, then $\sup_n f_n$, $\inf_n f_n$, $\limsup_{n\to\infty} f_n$ and $\liminf_{n\to\infty} f_n$ are measurable. But it should be observed that this desirable closure property does not persist for families (which, of course, may be uncountable) of measurable functions.

Given a measure μ, a property is said to hold **μ-almost everywhere** (abbreviated to μ-a.e.) in X if it holds in a set $X \setminus N$ where N is a set such that $\mu(N) = 0$.

There is a very surprising result about the uniform convergence of sequences of measurable functions. This is

THEOREM 1.11. (Egorov's Theorem). *Suppose that $\mu(X) < +\infty$, that $\{f_n\}$ is a sequence of functions each of which is measurable and is finite μ-a.e. on X, and that* $\lim_{n\to\infty} f_n(x)$ *exists μ-a.e. and is finite μ-a.e.*

Then, given $\varepsilon > 0$, there is a set $Q \subset X$ such that $f_n(x)$ converges uniformly on Q and $\mu(X \setminus Q) < \varepsilon$.

See, for instance, Saks (p. 18) for a proof.

It follows from Theorem 1.3 that semi-continuous functions are measurable.

For any measurable function f we define its integral with respect to a measure μ as follows:

First suppose that f is non-negative on X. Then say that the sets $\{A_k\}_{(k=1,\ldots,n)}$ are a partition of X if $\bigcup_{k=1}^{i} A_k = X$, $A_i \cap A_j = \phi$ if $i \neq j$ and $A_i \in \mathfrak{B}$. Let

$$S[f; \{A_k\}] = \sum_{k=1}^{n} \inf_{x \in A_k} f(x)\, \mu(A_k)$$

Then the integral of f over X with respect to μ, $\int f(x)\, \mu(dx)$, is defined by

$$\int f(x)\, \mu(dx) = \sup S[f; \{A_k\}],$$

the supremum being taken over all partitions of X. Naturally, we allow the integral to have the value ∞.

If, next, f may have arbitrary sign over X, we define

$$f_+(x) = \max[f(x), 0]; \qquad f_-(x) = -\min[f(x), 0],$$

so that

$$f(x) = f_+(x) - f_-(x); \qquad |f(x)| = f_+(x) + f_-(x)$$

and f_+, f_- are non-negative. Furthermore, f_+ and f_- are measurable, and so the integral of each with respect to μ is defined.

If not both $\int f_+(x)\,\mu(dx)$ and $\int f_-(x)\,\mu(dx)$ have the value ∞, we define the integral of f by

$$\int f(x)\,\mu(dx) = \int f_+(x)\,\mu(dx) - \int f_-(x)\,\mu(dx),$$

and if both are finite, we say that f is integrable with respect to μ. Clearly, f_+, f_- and $|f|$ are integrable with f.

We note that the above definition implicitly also defines the integral over any set $E \in \mathfrak{B}$ since E may be regarded as a locally compact space with the topology induced by that on X, and the Borel tribe on E is then induced by setting $\mathfrak{B}_E = \{B \cap E \mid B \in \mathfrak{B}\}$.

If $f(x) = g(x)$ μ-a.e. on X, then the integrals with respect to μ of f and g are equal. If f and g are integrable with respect to μ, so is $\alpha f + \beta g$, $\alpha, \beta \in \mathbb{R}$ and

$$\int (\alpha f + \beta g)(x)\,\mu(dx) = \alpha \int f(x)\,\mu(dx) + \beta \int g(x)\,\mu(dx).$$

Furthermore, if $A = \bigcup_{n=1}^{\infty} A_n$, where $\{A_n\}$ are disjoint Borel sets and f is μ-integrable over A, then

$$\int_A f(x)\,\mu(dx) = \sum_{n=1}^{\infty} \int_{A_n} f(x)\,\mu(dx).$$

Also, if $f(x) \leqslant g(x)$ in supp μ, then $\int f(x)\,\mu(dx \leqslant \int g(x)\,\mu(dx)$. There is also a slightly more precise result than this, which is of frequent use.

Lemma 1.12. *If $f(x) < g(x)$ in* supp μ, *then* $\int f(x)\,\mu(dx) < \int g(x)\,\mu(dx)$.

Let $S_n = \{x \in \operatorname{supp}\mu \mid f(x) \leqslant g(x) - 1/n\}$. Then $\bigcup_{n=1}^{\infty} S_n = \operatorname{supp}\mu$, and so $\mu(\bigcup_{n=1}^{\infty} S_n) > 0$. By (*ii*) it follows that there is an n such that $\mu(S_n) > 0$. Then

$$\int f(x)\,\mu(dx) = \int_{S_n} f(x)\,\mu(dx)$$
$$+ \int_{X \setminus S_n} f(x)\,\mu(dx) \leqslant \int_{S_n} (g(x) - 1/n)\,\mu(dx) + \int_{X \setminus S_n} g(x)\,\mu(dx)$$

so that

$$\int f(x)\,\mu(dx) \leqslant \int g(x)\,\mu(dx) - 1/n\,\mu(S_n) < \int g(x)\,\mu(dx).$$

Thus, in many respects, the integral behaves as, for example, the usual Riemann integral does, and it does so in respect of uniform convergence as well; for we have

THEOREM 1.13. *Suppose that* $\{f_n\}$ *is a sequence of functions, each of which is real-valued and integrable with respect to* μ *over a set* E *of finite* μ*-measure. Suppose further that* f_n *converges uniformly on* E *to a function* f. *Then* f *is integrable with respect to* μ *and*

$$\lim_{n\to\infty} \int f_n(x)\,\mu(dx) = \int \lim_{n\to\infty} f_n(x)\,\mu(dx).$$

But, in fact, more is true. Uniform convergence is not necessary here, as it almost always is in Riemann integration. We have

THEOREM 1.14 (Beppo Levi). *If* $\{f_n\}$ *is an increasing sequence of non-negative measurable functions then*

$$\lim_{n\to\infty} \int f_n(x)\,\mu(dx) = \int \lim_{n\to\infty} f_n(x)\,\mu(dx). \tag{1.4}$$

More generally, we have

THEOREM 1.15 (Fatou's Lemma). *Let* $\{f_n\}$ *be a sequence of non-negative measurable functions. Then*

$$\int \liminf_{n\to\infty} f_n(x)\,\mu(dx) \leqslant \liminf_{n\to\infty} \int f_n(x)\,\mu(dx)$$

as well as

THEOREM 1.16 (Lebesgue's Dominated Convergence Theorem). *If* $\{f_n\}$ *is a sequence of measurable functions which converges* μ*-a.e., and if there is an integrable function* g *such that*

$$|f_n(x)| \leqslant g(x) \quad \mu\text{-a.e. on } X,$$

then (1.4) *holds.*

For the proof of this and the previous theorem see Saks (pp. 28, 29).

It has already been pointed out that there is no satisfactory theory for families of measurable functions. If, however, we restrict ourselves to semi-continuous functions then something can be said. First, we need a lemma concerning l.s.c. functions:

LEMMA 1.17. *Suppose that* g *is l.s.c. on the compact set* K *and that*

$$\mathscr{F}(K) = \{f;\, K \to \mathbb{R} \mid f(x) \leqslant g(x) \text{ and } f \text{ continuous on } K\}.$$

Then

$$\int g(x)\,\mu(dx) = \sup\left\{\int f(x)\,\mu(dx) \mid f \in \mathscr{F}(K)\right\}.$$

Since g is measurable and is bounded below on K, its integral exists. Also, if $f \in \mathscr{F}(K)$, then $\int f(x)\,\mu(dx) \leqslant \int g(x)\,\mu(dx)$.

Since g is bounded below, we may, by adding a suitable constant, suppose that $g \geqslant 1$. Then, by Theorem 1.7(2), we may suppose that there is an increasing sequence of functions $\{f_n\}$, each of which is continuous and positive, which converges to g. By Theorem 1.14

$$\lim_{n\to\infty} \int f_n(x)\, \mu(dx) = \int g(x)\, \mu(dx);$$

and so, given $\varepsilon > 0$, there is an $f_n \in \mathscr{F}(K)$ such that

$$\int f_n(x)\, \mu(dx) > \int g(x)\, \mu(dx) - \varepsilon.$$

This now gives the result.

With this preliminary, albeit important, result established, we are in a position to prove an analogue of the Beppo Levi Theorem, in which an increasing sequence is replaced by an up-directed family.

THEOREM 1.18. *Let $\{g_\alpha\}_{\alpha\in A}$ be an up-directed family of functions, each of which is l.s.c. on the compact space X. Then*

$$\sup_{\alpha\in A} \int g_\alpha(x)\, \mu(dx) = \int \sup_{\alpha\in A} g_\alpha(x)\, \mu(dx). \tag{1.5}$$

We note first that the integrals all exist, since by Theorem 1.2 all the integrands on the left are bounded below, and by Theorem 1.6 that on the right is l.s.c. (and clearly bounded below).

Let $g = \sup_A g_\alpha$. Since $g_\alpha \leqslant g$, we have

$$\sup_A \int g_\alpha(x)\, \mu(dx) \leqslant \int g(x)\, \mu(dx),$$

and it remains to prove that

$$\sup_A \int g_\alpha(x)\, \mu(dx) \geqslant \int g(x)\, \mu(dx) \tag{*}$$

Suppose first that $g = 0$. Then, arguing as in the proof of Theorem 1.6, we have, given $\varepsilon > 0$, that there is $\alpha_0 \in A$, such that, for $x \in X$,

$$g_{\alpha_0}(x) > -\varepsilon/\mu(X)$$

Hence

$$\int g_{\alpha_0}(x)\, \mu\, (dx) > -\varepsilon,$$

and, since ε is arbitrary we have (*) in the special case in which $g = 0$. Turning to the general case, let $f \in \mathscr{F}(X)$ be such that

$$\int f(x)\, \mu(dx) > \int g(x)\, \mu(dx) - \varepsilon.$$

Now $\{-(g_\alpha - f)_-\}_{\alpha\in A}$ is an up-directed family of l.s.c. functions of which the supremum is 0, and so

$$\sup_A \{-\int (g_\alpha(x) - f(x)_-\, \mu(dx)\} = 0$$

Consequently $\sup_A \int (g_\alpha - f)(x)\,\mu(dx) \geqslant 0$; and thus

$$\sup_A \int g_\alpha(x)\,\mu(dx) \geqslant \int f(x)\,\mu(dx) > \int g(x)\,\mu(dx) - \varepsilon,$$

which gives (1.5).

It is now immediate that

THEOREM 1.19. *If $\{g_\alpha\}_{\alpha \in A}$ is a down-directed family of functions each of which is u.s.c. on the compact space X, then*

$$\inf_A \int g_\alpha(x)\,\mu(dx) = \int \inf_A g_\alpha(x)\,\mu(dx).$$

Suppose now that X is a locally compact Hausdorff Space. Let $\mathscr{C}(X)$ denote the set of all real-valued functions each of which has compact support and is continuous on X. Then $\mathscr{C}_0(X)$ is a linear space over $\mathbb{R}$.

Given any Radon measure μ on X, we define a linear functional ϕ_μ on $\mathscr{C}_0(X)$ by

$$\phi_\mu(f) = \int f(x)\,\mu(dx).$$

It is clear that ϕ_μ is a positive linear functional, i.e. if $f \geqslant 0$, then $\phi_\mu(f) \geqslant 0$ and ϕ_μ is linear. Thus, with each Radon measure we may associate a positive linear functional on $\mathscr{C}_0(X)$.

It is an extremely important result that the converse of this is true. Thus we have

THEOREM 1.20 (The Riesz Representation Theorem). *Let X be a locally compact Hausdorff space and let ϕ be a positive linear functional on $\mathscr{C}_0(X)$. Then there is one, and only one, Radon measure μ_ϕ such that*

$$\phi(f) = \int f(x)\,\mu_\phi\,(dx)$$

for all $f \in \mathscr{C}_0(X)$.

In fact, μ_ϕ is defined for any open set G by setting

$$\mu_\phi(G) = \sup\{\phi(f) \mid f \leqslant \chi_G\},$$

where χ_G is the indicator function for G, i.e. $\chi_G(x) = 1$ for $x \in G$ and $\chi_G(x) = 0$ for $x \notin G$. Then, for any set M, we set

$$\mu_\phi(M) = \inf\{\mu_\phi(G) \mid G \supset M\}.$$

It then turns out that μ_ϕ is a Radon measure. The reader is referred to Rudin (pp. 40-47) for the proof.

In some versions of integration theory (of which the Bourbaki account is perhaps the archetype), the result of Theorem 1.20 becomes the starting point. Thus, when X is compact, a measure on X is defined to be a positive linear functional μ on $\mathscr{C}(X)$, and the integral $\int f(x)\, \mu(dx)$ is, when $f \in \mathscr{C}(X)$, merely a synonym for $\mu(f)$. Then the content of Theorem 1.19 serves as the definition of the integral of an l.s.c. function, viz, if g is l.s.c. on X,

$$\int g(x)\, \mu(dx) = \sup \{\int f(x)\, \mu(dx) \mid f \in \mathscr{C}(X) \text{ and } f \leqslant g\}.$$

Likewise, if h is u.s.c. on X, then, by definition,

$$\int h(x)\, \mu(dx) = \inf \{\int f(x)\, \mu(dx) \mid f \in \mathscr{C}(X) \text{ and } f \geqslant g\}.$$

Then, given any positive numerically valued function f, we define the upper and lower integrals of f by

$$\overline{\int} f(x)\, \mu(dx) = \inf \{\int g(x)\, \mu(dx) \mid g \text{ l.s.c. and } g \geqslant f\}$$
$$\underline{\int} f(x)\, \mu(dx) = \sup \{\int h(x)\, \mu(dx) \mid h \text{ u.s.c. and } h \leqslant f\}.$$

There is here no question of f being measurable. The theory then proceeds by saying that a non-negative function is integrable if its upper and lower integrals coincide and their common value is finite.

We shall show that, if a non-negative function is measurable in the other sense we have already described, then its upper and lower integrals coincide.

LEMMA 1.21. *Suppose that f is μ-measurable and non-negative in the compact space E. Then, given $\varepsilon > 0$, there is an l.s.c. function g, such that $g \geqslant f$ and $\int\limits_E g(x)\, \mu(dx) \leqslant \int f(x)\, \mu(dx) + \varepsilon$.*

If $\int f(x)\, \mu(dx) = \infty$, there is nothing to prove. Suppose, then, that $\int f(x)\, \mu(dx) < \infty$. Let $\eta = \varepsilon/(1+\mu(E))$ and define E_k by

$$E_k = \{x \in E \mid (k-1)\eta \leqslant f(x) \leqslant k\eta\}.$$

Let $G_k \supset E_k$ be open and such that $\mu(G_k) \leqslant \mu(E_k) + 1/(k2^k)$. Let

$$g(x) = \sum_{k=1}^{\infty} k\eta\, \chi_k(x)$$

where χ_k is the indicator function for G_k. Then χ_k is l.s.c., and so, by Theorems 1.1 and 1.6, g is l.s.c. and, for $x \in X$ and each k

$$g(x) \geqslant k\eta\, \chi_k(x).$$

Hence $g(x) \geqslant k\eta \geqslant f(x)$ when $x \in E_k$, and so $g(x) \geqslant f(x)$ everywhere in E. Also

$$\begin{aligned} \int g(x)\, \mu(dx) &= \sum k\eta\, \mu(G_k) \\ &= \sum_{k=1}^{\infty} (k-1)\, \eta\, \mu(E_k) + \eta \sum_{k=1}^{\infty} \mu(E_k) + \sum_{k=1}^{\infty} k\eta\, \mu(G_k \setminus E_k), \end{aligned}$$

and this last is not greater than $\int f(x)\, \mu(dx)+\eta(1+\mu(E))$, which gives the result.

This lemma then expresses the fact that

$$\int f(x)\, \mu(dx) = \bar{\int} f(x)\, \mu(dx).$$

LEMMA 1.22. *Suppose that f is μ-measurable and non-negative in E. Then, given $\varepsilon > 0$, there is a u.s.c. function $h \leqslant f$* such that

$$\int h(x)\, \mu(dx) \geqslant \int f(x)\, \mu(dx) - \varepsilon.$$

Suppose first that f is bounded. Then $f^{-1} \geqslant (\sup f)^{-1}$ and is non-negative and measurable. By Lemma 1.21 we can find an l.s.c. function g such that $g \geqslant 1/f$ and $\int (g(x)-1/f(x))\, \mu(dx) < \varepsilon/(\sup f)^2$. Now $1/g$ is u.s.c. and $1/g \leqslant f \leqslant \sup f$. Also

$$\int (f(x)-1/g(x))\, \mu(dx) \leqslant \sup f \int [f(x)\, g(x)-1]\, \mu(dx)$$
$$\leqslant (\sup f)^2 \int [g(x)-1/f(x)]\, \mu(dx)$$

and this last is less than ε.

Thus we can find a u.s.c. function $h \leqslant f$, such that

$$\int h(x)\, \mu(dx) \geqslant \int f(x)\, \mu(dx) - \varepsilon.$$

Now suppose that f is integrable, and let $f_n(x) = \min [f(x), n]$. Then f_n is bounded, and there is an u.s.c. function $h_n \leqslant f_n$ such that

$$\int (f_n(x) - h_n(x))\, \mu(dx) < \tfrac{1}{2}\varepsilon.$$

Also we can find n such that $\int [f(x) - f_n(x)]\, \mu(dx) < \frac{1}{2}\varepsilon$ and, for this n

$$\int (f(x) - h_n(x))\, \mu(dx) < \varepsilon.$$

If f is not integrable, then $\int f_n(x)\, \mu(dx)$ diverges from ∞ as n tends to ∞, and so therefore does $\int h_n(x)\, \mu(dx)$.

This lemma then expresses the fact that

$$\int f(x)\, \mu(dx) = \underline{\int} f(x)\, \mu(dx).$$

and thus, if f is integrable its upper and lower integrals coincide.

Given two locally compact Hausdorff spaces X and Y, and given that λ and μ are Radon measures defined on X and Y respectively, the product measure $\lambda \times \mu$, defined on $X \times Y$, is arrived at in the following way:

Suppose that $K \subset X$, $L \subset Y$ are compacts, and that $f(x, y)$ is continuous on $X \times Y$ and $\operatorname{supp} f \subset K \times L$. Then it may be shown that

$$h(y) = \int f(x, y)\, \lambda(dx)$$

is continuous in Y and that $\operatorname{supp} h \subset L$.

Then $\lambda \times \mu$ is defined as a linear functional on $\mathscr{C}_0(X \times Y)$ by

$$(\lambda \times \mu)\,(f) = \mu(h).$$

Since it is a positive functional, it has, by Theorem 1.20, a measure associated with it, and this is the product measure $\lambda \times \mu$ of λ and μ.

Furthermore, we have Fubini's Theorem, that

$$\int f(x, y)\, (\lambda \times \mu)\, [d(x, y)] = \iint f(x, y)\, \mu(dy)\, \lambda(dx)$$
$$= \iint f(x, y)\, \lambda(dx)\, \mu(dy)$$

for any function integrable in $X \times Y$. The result is also true without qualification when f is non-negative, if it is measurable. Reference may be made to Bourbaki [1] for details.

An alternative, and in many ways more attractive, treatment in measure theoretic terms is to be found in Rudin. However, it is then necessary to introduce the restriction that the measures concerned be σ-finite.

§ 1.3. The Space $\mathbb{R}^p$

$\mathbb{R}^p$ is, of course, the metric space of which the points x are p-tuples $(x^1, x^2, \ldots, x^p)$ of real numbers, and which is provided with a metric d given by

$$d(x, y) = \sqrt{\{\sum_{r=1}^{p} (x^r - y^r)^2\}}.$$

$\mathbb{R}^p$ is also a linear space over $\mathbb{R}$, so that expressions like $x+y$, λx and $x-y$ are well defined. We shall usually denote $d(x, y)$ by $|x-y|$. $\mathbb{R}^p$ is locally compact and Hausdorff and, in $\mathbb{R}^p$, a set is compact if, and only if, it is closed and bounded.

Among all the measures on $\mathbb{R}^p$ one has special importance. This is **Lebesgue p-measure.** It can be defined in the following way:

Given any $f \in \mathscr{C}_0(\mathbb{R}^p)$, the Riemann integral $\int_{\mathbb{R}^p} f(x)\, dx$ is well defined. There is thus defined on $\mathscr{C}_0(\mathbb{R}^p)$ a positive linear functional. By Theorem 1.20 this gives rise to a Radon measure, and this measure is said to be Lebesgue p-measure, which we usually denote by m.

Since $\int_{\mathbb{R}^p} f(x+a)\, dx = \int_{\mathbb{R}^p} f(x)\, dx$, it follows immediately that m is **translation invariant,** i.e. for $E \in \mathfrak{B}(\mathbb{R}^p)$,

$$m(a+E) = m(E).$$

It can also be shown that any Radon measure on $\mathbb{R}^p$ which is translation invariant is a positive multiple of m.

We shall denote the integral $\int f(x)\, m(dx)$ by $\int f(x)\, dx$, and will say 'integrable' when m-integrable is meant, and 'a.e.' for m-a.e.

Because, in $\mathbb{R}^p$, continuous limits and sequential limits are equivalent, we can in this context enunciate a continuous version of Theorem 1.16.

THEOREM 1.23. *Let X be a locally compact Hausdorff space and let $G \subset \mathbb{R}^p$ be open. Suppose that μ is a Radon measure on X, and that*

$$f : X \times G \to \overline{\mathbb{R}}$$

is such that

(*i*) *$f(x, y)$ is measurable on X for each $y \in G$;*

(*ii*) *$\lim_{y \to a} f(x, y)$ exists μ-a.e. in X;*

(*iii*) *There is a μ-integrable function g such that, for $y \in G$, $|f(x, y)| < g(x)$ μ-a.e. in X.*

Then, for $a \in G$,

$$\lim_{y \to a} \int f(x, y)\, \mu(dx) = \int \lim_{y \to a} f(x, y)\, \mu(dx).$$

There is also a continuous version of Fatou's Lemma, which takes the following form:

THEOREM 1.24. *Let X, G and μ be as in Theorem 1.23. Suppose that $f(x, y)$ is measurable and non-negative for each $y \in G$. Then, for $a \in G$,*

$$\int \liminf_{y \to a} f(x, y)\, \mu\,(dx) = \liminf_{y \to a} \int f(x, y)\, \mu\,(dx).$$

To prove these two theorems we need only remark that '$\lim_{y \to a} f(x, y)$ exists' is equivalent to '$\lim_{y \to \infty} f(x, y_n)$ exists for each sequence converging to a' and apply Theorems 1.15 and 1.16.

In $\mathbb{R}^p$ also, differentiation is a meaningful process, and we may use Theorem 1.23 to prove a theorem of considerable generality concerning differentiation under the integral sign.

THEOREM 1.24. *Let X and G be as in Theorem 1.22. Suppose that μ is a Radon measure on X and that*

$$f : \; X \times G \to \mathbb{R}$$

is such that

(*i*) *$f(x, y)$ is integrable in X for each $y \in G$;*

(*ii*) *$\dfrac{\partial}{\partial y^r}\{f(x, y)\}$ exists for each $y \in G$ and μ-a.e. in X;*

(*iii*) *there is a μ-integrable function g such that*

$$\left| \frac{\partial}{\partial y^r}\{f(x, y)\} \right| \leqslant g(x) \text{ for } y \in G \text{ and } \mu\text{-a.e. in } X.$$

Then, for each $y \in G$,

$$\frac{\partial}{\partial y^r}\{\int f(x, y)\, \mu(dx)\} = \int f_{y^r}(x, y)\, \mu(dx).$$

Let $a \in G$, and let $h = (0, \ldots, 0, \lambda, 0, \ldots, 0)$ where λ is the r-th coordinate. Then

$$\lim_{h\to 0} \frac{f(x, a+h)-f(x, a)}{h} \text{ exists } \mu\text{-a.e. in } X$$

and

$$\left|\frac{f(x, a+h)-f(x. a)}{h}\right| = |f_{y^r}(x, a+\theta h)| \leqslant g(x)$$

for sufficiently small h; so, by Theorem 1.23, letting $F(y) = \int f(x,y)\, \mu(dx)$, we have

$$\lim_{h\to 0} \frac{F(a+h)-F(a)}{h} = \int f_{y^r}(x, a)\, \mu(dx),$$

showing that $F_{y^r}(a)$ exists and has the stated value.

By a repeated application of Fubini's Theorem we may show that, for any f which is Lebesgue integrable over $\mathbb{R}^p$,

$$\int f(x)dx = \int_{-\infty}^{\infty} \ldots \int_{-\infty}^{\infty} f(x^1, \ldots, x^p)\, dx^1 \ldots dx^p.$$

The following theorem is fundamental for any discussion of change of variable.

THEOREM 1.25. *Let* $G \subset \mathbb{R}^p$ *be open and let* $h: G \to \mathbb{R}^p$ *be a mapping which has continuous first partial derivatives. Suppose that*

$$h(\xi) = (h^1(\xi), h^2(\xi), \ldots, h^p(\xi))$$

and that $J(\xi) = \det\left[\frac{\partial h^i(\xi)}{\partial \xi^j}\right] \neq 0$ *in* G.

Then the function $f(x)$ *is integrable over* $h(G)$ *if, and only if,* $f(h(\xi)\, J(\xi)$ *is integrable over* G *and*

$$\int_{h(G)} f(x)dx = \int_G f(h(\xi))\, |J(\xi)|\, d\xi.$$

See Zaanen (pp. 162–167) for the proof.

Let, in particular,

$$G = \{(\rho, \theta_1, \ldots, \theta_{p-1}) \mid 0 < \rho < \infty, 0 < \theta_i < \pi\ (i = 1, \ldots, p-2) \text{ and } 0 \leqslant \theta_{p-1} \leqslant 2\pi\}$$

and let $x = h(\xi)$ be given by

$$x^i = \rho \sin \theta_1 \dots \sin \theta_{i-1} \cos \theta_i \ (i = 1, \dots, p-1)$$
$$x^p = \rho \sin \theta_1 \dots \sin \theta_{p-2} \sin \theta_{p-1} \qquad (1.6)$$

Then $h(G) = \mathbb{R}^p \setminus \{x \in \mathbb{R}^p \mid x^{p-1} = 0 \text{ and } x^p = 0\}$ and

$$J(\xi) = \rho^{p-1} \sin^{p-2} \theta_1 \dots \sin \theta_{p-2},$$

and so $f(x)$ is integrable over $h(G)$ if, and only if, $f(h(\xi)J(\xi)$ is integrable over G, and we have equality of the integrals.

Furthermore, $\mathbb{R}^p \setminus h(G)$ and

$$\{(\rho_1\, \theta_1, \dots, \theta_{p-1}) \mid 0 \leqslant \rho < \infty, 0 \leqslant \theta_i \leqslant \pi \ (i = 1, 2, \dots, p-2) \text{ and } 0 \leqslant \theta_{p-1} \leqslant 2\pi\} \setminus G$$

are of zero Lebesgue p-measure. So, using Fubini's Theorem, we have

$$\int_{-\infty}^{\infty} \dots \int_{-\infty}^{\infty} f(x^1, \dots, x^p)\, dx^1 \dots dx^p$$

$$= \int_0^{\infty} \int_0^{2\pi} \int_0^{\pi} \dots \int_0^{\pi} F(\rho, \theta_1, \dots, \theta_{p-1})\, \rho^{p-1} \sin^{p-2} \theta_1 \dots \sin \theta_{p-2}$$
$$d\theta_1 \dots d\theta_{p-1}\, d\rho,$$

where $F(\rho, \theta_1, \dots, \theta_{p-1}) = f(h(\xi))$.

We now define two other particular measures which we shall be using constantly in later chapters. First, **the ball measures.** We define the open ball of radius r and centre a to be

$$B_r(a) = \{x \in \mathbb{R}^p \mid |x-a| < r\},$$

and the annulus $B_{r,R}(a)$ of centre a by

$$B_{r,R}(a) = \{x \in \mathbb{R}^p \mid r < |x-a| < R\}.$$

By taking $f = \chi_{B_r(a)}$ and making the substitution given by replacing x^i by $x^i - a^i$ on the left of (6) we have

$$mB_r(a) = \int_0^{r} \int_0^{2\pi} \int_0^{\pi} \dots \int_0^{\pi} \rho^{p-1} \sin^{p-2} \theta_1 \dots \sin \theta_{p-2}\, d\theta_1 \dots d\theta_{p-1}\, d\rho$$

$$= \frac{\pi^{\frac{1}{2}p}\, r^p}{\Gamma(\frac{1}{2}p+1)}$$

Similarly,

$$mB_{r,R}(a) = \frac{\pi^{\frac{1}{2}p}}{\Gamma(\frac{1}{2}p+1)} (R^p - r^p).$$

The ball measure β_r^a is defined on $\mathbb{R}^p$ by

$$\beta_r^a(E) = \frac{\Gamma(\frac{1}{2}p+1)}{\pi^{\frac{1}{2}p}\, r^p}\, m\,(E \cap B_r(a)) \tag{1.7}$$

for every Borel set $E \subset \mathbb{R}^p$. Then $\beta_r^a(\mathbb{R}^p) = 1$ and β_r^a is supported on $\bar{B}_r(a)$, i.e. supp $\beta_r^a = \bar{B}_r(a)$.

The annulus measure $\beta_{r,R}^a$ is similarly defined by

$$(R^p - r^p)\, \beta_{r,R}^a(E) = \frac{\Gamma(\frac{1}{2}p+1)}{\pi^{\frac{1}{2}p}}\, m\,(E \cap B_{r,R}(a))$$

and then $\beta_{r,R}^a\,(\mathbb{R}^p) = 1$ and $\beta_{r,R}^a$ is supported on $\bar{B}_{r,R}\,(a)$.

Clearly $\beta_r^a(a+E) = \beta_r^0(E)$, and so

$$\int f(t)\, \beta_r^a(dt) = \int f(a+t)\, \beta_r^0(dt)$$

and there is a similar result for annulus measure.

A function f is said to be locally integrable in an open set G if it is integrable in every compact subset of G. For a locally integrable f we define the **ball mean** $\mathscr{B}_r^a(f)$ by

$$\mathscr{B}_r^a(f) = \int f(t)\, \beta_r^a(dt).$$

The sphere $S_r(a)$ of radius r and centre a is defined to be

$$S_r(a) = \{x \in \mathbb{R}^p \mid |x-a| = r\}$$

Given a function f defined on $\mathbb{R}^p$, let the restriction of f to $S_r(a)$ be denoted by f_r, and define the function $F(\rho, \theta_1, \ldots, \theta_{p-1})$ by

$$F(\rho, \theta_1, \ldots, \theta_{p-1}) = f_r\,(x^1, \ldots, x^p)$$

where $\rho, \theta_1, \ldots \theta_{p-1}$ and $x^1, \ldots, x^p$ are related by the variant of (1.6) used a moment ago.

Let ϕ_r^a be defined on $\mathscr{C}_0(\mathbb{R}^p)$ by

$$\phi_r^a(f) = \frac{\Gamma(\frac{1}{2}p)}{2\pi^{\frac{1}{2}p}} \int_0^{2\pi}\int_0^{\pi} \cdots \int_0^{\pi} F(r, \theta_1, \ldots, \theta_{p-1})\, \sin^{p-2}\theta_1 \ldots \sin\theta_{p-2}\, d\theta_1 \ldots d\theta_{p-1} \tag{1.8}$$

Then ϕ_r^a is clearly a positive linear functional on $\mathscr{C}_0(\mathbb{R}^p)$, $\mathbb{R}^p$ is locally compact and so, by Theorem 1.20, ϕ_r^a determines a unique Radon measure σ_r^a such that

$$\phi_r^a(f) = \int f(x)\, \sigma_r^a(dx)$$

This measure is called the **sphere measure*** of radius r and centre a. It is supported on $S_r(a)$ and $\sigma_r^a(\mathbb{R}^p) = 1$. Clearly $\sigma_r^a(a+E) = \sigma_r^0(E)$ and so

$$\int f(t)\, \sigma_r^a(dt) = \int f(a+t)\, \sigma_r^0(dt).$$

We define the **sphere mean** $\mathscr{S}_r^a(f)$ of a σ_r^a-integrable function f by

$$\mathscr{S}_r^a(f) = \int f(t)\, \sigma_r^a(dt).$$

There is then a useful relation between ball and sphere means. We have

$$\mathscr{B}_r^a(f) = \int f(x)\, \beta_r^a(dx) = \frac{\frac{p}{2}\Gamma\frac{p}{(2)}}{\pi^{\frac{1}{2}p} r^p} \int_0^r \int_0^{2\pi} \dots \int_0^{\pi} F(\rho, \theta_1, \dots, \theta_{p-1})$$

$$\rho^{p-1} \sin^{p-2}\theta_1 \dots \sin\theta_{p-2}\, d\theta_1 \dots d\theta_{p-1}\, d\rho$$

and, using Fubini's Theorem and (1.8), this last is equal to

$$\frac{p}{r^p} \int_0^r \rho^{p-1}\, d\rho \int f(x)\, \sigma_\rho^a(dx),$$

first for any continuous function, and then by extension for any semi-continuous function. Thus, for such functions

$$r^p\, \mathscr{B}_r^a(f) = p \int_0^r \rho^{p-1}\, \mathscr{S}_\rho^a(f) d\rho \tag{1.9}$$

Similarly

$$(R^p - r^p)\, \mathscr{B}_{r,R}^a(f) = p \int_r^R \rho^{p-1}\, \mathscr{S}_\beta^a(f) d\rho \tag{1.10}$$

where $\mathscr{B}_{r,R}^a$ denotes the obvious **annulus mean** analogous to the ball mean.

Finally, from known results about the differentiability of additive set functions (Saks, p. 118, Theorem 6.3) we may conclude that

THEOREM 1.27. *If f is integrable over an open set $G \subset \mathbb{R}^p$ then*

$$\lim_{r\to 0} (mB_r(a))^{-1} \int_{B_r(a)} f(x)dx = f(a) \text{ a.e. in } G.$$

We may rephrase this result as

$$\int f(x)\, \beta_r^a(dx) = f(a) + o(1), \text{ as } r \to 0 \text{ a.e. in } G.$$

* An alternative description of σ_r^a is as follows: Let $M \in S_r(a)$, and let

$$C(M) = \{(1-\lambda)a + \lambda\xi \mid 0 < \lambda < 1 \text{ and } \xi \in M\}.$$

Then, for any Borel set E,

$$\sigma_r^a(E) = \frac{\Gamma\left(\frac{p}{2}+1\right)}{\Pi^{\frac{1}{2}p} r^p}\, m\, C(E \cap S_r(a)).$$

§ 1.4. Approximation by Smooth Functions

Let $G \subset \mathbb{R}^p$ be open. Then a function ϕ, defined in G such that supp $\phi \subset G$ and ϕ is infinitely differentiable, is said to be a **test function** in G. Let $\mathscr{D}(G)$ denote the set of all such test functions. It is not immediately clear, as Laurent Schwartz himself has remarked, that $\mathscr{D}(G)$ is non-void; but, in fact, there is a plentiful supply of test functions in G. To see that this is so, consider the function τ_δ defined on $\mathbb{R}^p$ by

$$\tau_\delta(x) = C_\delta e^{-1/(\delta^2 - |x|^2)} \text{ for } |x| < \delta$$
$$= 0 \text{ for } |x| \geqslant \delta.$$

and with C_δ so chosen that $\int_{\mathbb{R}^p} \tau_\delta(x)dx = 1$.

It is of interest to find an exact order of magnitude for C_δ when δ is small. The surprising thing is that C_δ is so very big.

LEMMA 1.28. *Let $0 < \delta \leqslant 1$. Then C_δ lies between constant positive multiples of $\delta^{-2p} e^{1/\delta^2}$.*

Transforming to spherical polar coordinates as given by (1.6) we have

$$\int_{|t|<\delta} e^{-1/(\delta^2 - |t|^2)} dt = \frac{2\pi^{\frac{1}{2}p}}{\Gamma(\frac{1}{2}p)} \int_0^\delta e^{-1/(\delta^2 - r^2)} r^{p-1} dr.$$

In the right hand integral set $\delta^2 - r^2 = \delta^2/(1+\delta^2 s)$. Then the right hand side equals

$$\frac{\pi^{\frac{1}{2}p}}{\Gamma(\frac{1}{2}p)} \delta^{2p} e^{-1/\delta^2} \int_0^\infty s^{\frac{1}{2}p-1} (1+\delta^2 s)^{-\frac{1}{2}p-1} e^{-s} ds.$$

This last integral is less than $\int_0^\infty s^{\frac{1}{2}p-1} e^{-s} ds = \Gamma(\frac{1}{2}p)$, and so

$$\int_{|t|<\delta} e^{-1/(\delta^2 - |t|^2} dt < \pi^{\frac{1}{2}p} \delta^{2p} e^{-1/\delta^2}.$$

On the other hand, the last integral is greater than that over the range $[0, 1]$, and this is greater than

$$(1+\delta^2)^{-\frac{1}{2}p-1} \int_0^1 s^{\frac{1}{2}p-1} e^{-s} ds \geqslant 2^{-\frac{1}{2}p-1} \int_0^1 s^{\frac{1}{2}p-1} e^{-s} ds$$

so that, setting $C = 2^{\frac{1}{2}p+1}(\int_0^1 s^{\frac{1}{2}p-1} e^{-s} ds)^{-1}$, we have

$$\pi^{-\frac{1}{2}p} \delta^{-2p} e^{1/\delta^2} < C_\delta < C\delta^{-2p} e^{1/\delta^2}.$$

Since the function $\psi(t)$, given by $\psi(t) = e^{-(1/t)}$ for $t > 0$ and $\psi(0) = 0$ for $t \leqslant 0$, is infinitely differentiable in $\mathbb{R}$, it follows that τ_δ is infinitely differentiable in $\mathbb{R}^p$. Furthermore, supp $\tau_\delta = \bar{B}_\delta(0)$ and so, given any open set $G \subset \mathbb{R}^p$, there is an $a \in G$ and $\delta > 0$ such that $\bar{B}_\delta(a) \subset G$, and so the function given by $\phi(x) = \tau_\delta(x-a)$ is a member of $\mathscr{D}(G)$.

Furthermore, $\mathscr{D}(G)$ is a linear subspace of $\mathscr{C}_0(G)$ which, we shall see in a moment, is 'dense' in $\mathscr{C}_0(G)$.

Let f be locally integrable in G, and let the convolution of f and τ_δ, $f*\tau_\delta$ be given by

$$(f*\tau_\delta)(x) = \int f(t)\,\tau_\delta(x-t)\,dt.$$

Then

THEOREM 1.29. *(i) $f*\tau_\delta$ is infinitely differentiable in G_δ where*

$$G_\delta = \{x \in G \mid \operatorname{dist}(x, \complement G > \delta)\},$$

and, if supp f is compact, $f*\tau_\delta \in \mathscr{D}(G)$ for sufficiently small δ.

(ii) If f is continuous in G then $f*\tau_\delta$ converges uniformly to f as $\delta \to 0$ in every compact subset of G.

(iii) $f*\tau_\delta$ converges to f as $\delta \to 0$ a.e. in G.

(iv)

$$\lim_{\delta \to 0} \int_K |(f*\tau_\delta)(x) - f(x)|\,dx = 0 \tag{1.11}$$

whenever $K \subset G$ is compact.

$f*\tau_\delta$ will be defined in G_δ, and repeated application of Theorem 1.24 ensures its infinite differentiability. Also, letting $K = \operatorname{supp} f$, when $\operatorname{dist}(x, K) > \delta$, then $B_\delta(x) \cap K = \phi$ and so, for $t \in \mathbb{R}^p, f(t)\tau_\delta(x-t) = 0$. Consequently, for these x, $(f*\tau_\delta)(x) = 0$; and so the set $\{x \in \mathbb{R}^p \mid \operatorname{dist}(x, K) \leqslant \delta\}$, which is compact, contains supp $(f*\tau_\delta)$. This gives (*i*) when δ is sufficiently small.

Turning to the proof of (*ii*), suppose that $a \in G$ and $B_{\delta_0}(a) \subset G$. Then $f*\tau_\delta$ is infinitely differentiable at a whenever $\delta < \delta_0$.

Given $\varepsilon > 0$, there is a $\lambda > 0$ such that, for $|t-a| < \lambda$, $|f(t) - f(a)| < \varepsilon$. Now

$$\begin{aligned}(f*\tau_\delta)(a) - f(a) &= \int (f(t) - f(a))\,\tau_\delta(x-t)\,dt \\ &= \int_{B_\delta(0)} (f(a-t) - f(a))\,\tau_\delta(t)\,dt.\end{aligned}$$

If $\delta < \lambda$ this integral is, in absolute value, not greater than $\varepsilon \int \tau_\delta(t)dt = \varepsilon$; and, since, given a compact K, λ will depend on K only, (*ii*) follows.

$$\begin{aligned}(iii)\ (f*\tau_\delta)(x) &= \int f(t)\,\tau_\delta(x-t)dt = \int f(x+t)\,\tau_\delta(t)dt \\ &= \frac{2\pi^{\frac{1}{2}p}}{\Gamma(\frac{1}{2}p)} \int_0^\delta \rho^{p-1}\,\tau_\delta(\rho) \int f(x+t)\,\sigma_\rho^0(dt)d\rho,\end{aligned}$$

where, by an abuse of notation, we denote $\tau_\delta(t)$ by $\tau_\delta(\rho)$ when $\rho = |t|$.

Thus

$$(f*\tau_\delta)(x) = \frac{2\pi^{\frac{1}{2}p}}{\Gamma(\frac{1}{2}p)}\int_0^\delta \rho^{p-1}\,\tau_\delta(\rho)\,\mathscr{S}_\rho^x(f)\,d\rho.$$

Let
$$g(r) = \int_{B_r(0)} f(x+t)dt = \frac{2\pi^{\frac{1}{2}p}}{\Gamma(\frac{1}{2}p)}\int_0^r \mathscr{S}_\rho^x(f)\,\rho^{p-1}\,d\rho.$$

Then, a.e. in G,
$$g(r) = f(x)\frac{\pi^{\frac{1}{2}p}r^p}{\Gamma(\frac{1}{2}p+1)}+\varepsilon(r)r^p$$

where
$$\varepsilon(r)\to 0 \text{ as } r\to 0.$$

Hence
$$(f*\tau_\delta)(x) = \int_0^\delta \tau_\delta(\rho)\,dg(\rho) = -\int_0^\delta g(\rho)\,\tau_\delta'(\rho)\,d\rho$$

$$= -\frac{\pi^{\frac{1}{2}p}}{\Gamma(\frac{1}{2}p+1)}f(x)\int_\delta^0 \rho^p\,\tau_\delta'(\rho)\,d\rho - \int_\delta^0 \varepsilon(\rho)\,\rho^p\,\tau_\delta'(\rho)\,d\rho$$

and

$$-\int \rho^p\,\tau_\delta'(\rho)\,d\rho = p\int_0^\delta \tau_\delta(\rho)\,\rho^{p-1}\,d\rho = p\,\frac{\Gamma(\frac{1}{2}p)}{2\pi^{\frac{1}{2}p}}\int \tau_\delta(x)dx$$

and this last has the value $\Gamma(\frac{1}{2}p+1)/\pi^{\frac{1}{2}p}$.

Furthermore $\int_0^\delta \varepsilon(\rho)\,\rho^{p-1}\,\tau_\delta'(\rho)\,d\rho\to 0$ as $\delta\to 0$,
and hence $(f*\tau_\delta)(x) = f(x)+0(1)$ as $\delta\to 0$,
which gives (*iii*).

(*iv*) Let $f = f_+ - f_-$. Then both f_+ and f_- are locally integrable and it is enough to show that (1.11) holds when $f\geqslant 0$. Assume this to be the case, and let $f_n(x) = \min\,[n, f(x)]$ so that f_n converges increasingly to f and thus, by Beppo Levi,

$$\int_k (f-f_n)(x)\,dx\to 0 \text{ as } n\to\infty.$$

Thus, given $\varepsilon > 0$, there is a bounded measurable $g\geqslant 0$ such that

$$\int_k |\,f(x)-g(x)\,|\,dx < \tfrac{1}{3}\varepsilon.$$

Let $\delta_n\to 0$. Then $g*\tau_{\delta_n}\to g$ a.e. and, by Egorov's Theorem (1.11.) there is a set $Q\subset K$ such that

$$m(K\setminus Q) < \varepsilon/3\,\sup_K g\,.\,mK, \text{ and } g*\tau_{\delta_n}\to g \text{ uniformly in } Q.$$

Now

$$\int_K |\,(g*\tau_{\delta_n})(x)-g(x)\,|\,dx = \Big(\int_Q+\int_{K\setminus Q}\Big)\,|\,(g*\tau_{\delta_n})(x)-g(x)\,|\,dx.$$

Choose N so that, for $n > N$, the first term is less than $\frac{1}{3}\varepsilon$, as we may by Theorem 1.13. The second term is less than

$$\int_{K \setminus Q} (g*\tau_{\delta_n})(x)\, dx + \int_{K \setminus Q} g(x)\, dx.$$

The second of these is less than $\sup g \,.\, m(K \setminus Q) < \frac{1}{3}\varepsilon$, while the first may be rewritten as

$$\int_{K \setminus Q} \int g(x-t)\, \tau_{\delta_n}(t)\, dt,$$

which is not greater than

$$\int_{K \setminus Q} \sup g \int \tau_{\delta_n}(t) dt = \sup g \,.\, m(K \setminus Q) < \tfrac{1}{3}\varepsilon;$$

so that
$$\int_K |(g*\tau_{\delta_n})(x) - g(x)|\, dx \to 0 \text{ as } n \to \infty;$$

whence
$$\int_K |(g*\tau_\delta)(x) - g(x)|\, dx \to 0 \text{ as } \delta \to 0.$$

Finally
$$\int_K |(f*\tau_\delta)(x) - f(x)|\, dx$$

is not greater than

$$\int_k |(f*\tau_\delta)(x) - (g*\tau_\delta)(x)|\, dx + \int_k |(g*\tau_\delta)(x) - g(x)|\, dx +$$
$$\int_k |g(x) - f(x)|\, dx$$

and the second term can be made smaller than $\frac{1}{3}\varepsilon$ by taking δ sufficiently small. The third term is less than $\frac{1}{3}\varepsilon$. The first term does not exceed

$$\int \int |f(t) - g(t)|\, \tau_\delta(x-t)\, dx dt$$

which is less than $\frac{1}{3}\varepsilon$. This now gives (1.11).

We may note, in conclusion that, if f is positive, then the approximating test functions to f given by this theorem are also positive, and that, if f is of compact support, then the approximating functions are of uniformly compact support, that is to say there is one compact set which contains the supports of all the approximating functions.

Theorem 1.29 expresses the fact that $\mathscr{D}(G)$, a linear subspace of both $\mathscr{C}_0(G)$ and $L^{\text{loc}}(G)$ (the space of all locally integrable functions in G), is, in the appropriate (but different) sense, 'dense' in each of these latter.

We have seen in Theorem 1.20 that a positive linear functional in $\mathscr{C}_0(X)$, with X a locally compact Hausdorff space, can be regarded as a measure. Thus a positive linear functional on $\mathscr{C}_0(G)$, with $G \subset \mathbb{R}^p$ open, 'is' a measure. We now show that, in the same sense, a positive linear functional on $\mathscr{D}(G)$ 'is' a measure.

THEOREM 1.30. *Let μ be a positive linear functional on $\mathscr{D}(G)$. Then there is one, and only one, positive linear functional μ^* on $\mathscr{C}_0(G)$ such that $\mu^*(\phi) = \mu(\phi)$ for all $\phi \in \mathscr{D}(G)$.*

Let μ be a positive linear functional on $\mathscr{D}(G)$. Let $\{\phi_n\}$, $\phi_n \in \mathscr{D}(G)$, be such that

(*a*) supp $\phi_n \subset K$ for some compact K independent of n.

(*b*) $\phi_m(x) - \phi_n(x) \to 0$ uniformly in G as $m, n \to \infty$.

Let $\theta_{nm}(x) = \phi_n(x) - \phi_m(x)$. By Urysohn's Lemma there is a non-negative continuous function of compact support taking the value 1 on the closure of an open set containing K. Then, as we may easily verify, the convolution of this function with τ_δ takes, for sufficiently small δ, the value 1 on K. Thus there is a non-negative $\chi \in \mathscr{D}(G)$ such that $\chi(x) = 1$ on K. Then, by (*b*), given $\varepsilon > 0$, there is an N such that, for $n > N$ and $m > N$,

$$| \theta_{nm}(x) | < \varepsilon;$$

and, by (*a*), when $n, m > N$, we have

$$-\varepsilon\, \chi(x) \leqslant \theta_{nm} < \varepsilon\, \chi(x)$$

Since μ is positive we now have

$$-\varepsilon\mu(\chi) < \mu(\theta_{nm}) = \mu(\phi_n) - \mu(\phi_m) < \varepsilon\mu(\chi)$$

whence, by Cauchy's Convergence criterion, $\lim_{n\to\infty} \mu(\phi_n)$ exists.

Thus for any $\{\phi_n\}$ satisfying (*a*) and (*b*), $\mu(\phi_n)$ converges.

Now suppose that $\psi \in \mathscr{C}_0(G)$. Then, by Theorem 1.29, there is a sequence $\{\phi_n\}$ satisfying (*a*) such that $\phi_n(x) \to \psi(x)$ uniformly. *Moreover,* $\lim_{n\to\infty} \mu(\phi_n)$ *depends only on* ψ.

For, suppose that $\{\phi_n\}$ satisfies (*a*) and $\lim_{n\to\infty} \phi_n'(x) = \psi(x)$. Then $\mu(\phi_n')$ converges and $\mu(\phi_n - \phi_n') = \mu(\phi_n) - \mu(\phi_n')$. Arguing as above, we may show that $\mu(\phi_n - \phi_n) \to 0$ as $n \to \infty$, so that $\lim_{n\to\infty} \mu(\phi_n') = \lim_{n\to\infty} \mu(\phi_n)$.

Thus, without ambiguity, we may define μ^* in $\mathscr{C}_0(G)$ by

$$\mu^*(\psi) = \lim_{n\to\infty} \mu(\phi_n)$$

We have remarked that when $\psi \geqslant 0$ we may assume its approximating sequence to be such that $\phi_n \geqslant 0$ and so μ^* is positive. Clearly μ^* is linear, and, since $\mu^*(\phi) = \mu(\phi)$ whenever $\phi \in \mathscr{D}(G)$, Theorem 1.30 follows.

§ 1.5. Vague Convergence

Suppose that a sequence $\{\mu_n\}$ of measures is given and that there is a measure μ such that

$$\lim_{n\to\infty} \mu_n(\phi) = \mu(\phi),\ \phi \in \mathscr{C}_0(\mathbb{R}^p)$$

Then the sequence $\{\mu_n\}$ is said to converge vaguely to μ and we write $\mu_n \rightsquigarrow \mu$.

The way in which the unique measure associated with the positive linear functional is defined in Theorem 1.20 makes it seem very plausible that an equivalent definition of vague convergence is that

$$\lim_{n\to\infty} \mu_n(E) = \mu(E), \ E \text{ a Borel set in } \mathbb{R}^p,$$

and this is, in fact, true.

We will collect here a number of results about vague convergence that will be utilised in Chapter 3.

LEMMA 1.31. *Suppose that $\mathscr{M}$ is a dense subset of $\mathscr{C}_0(\mathbb{R}^p)$. Suppose that $\{\mu_n\}$ is a sequence of measures and μ a measure such that*

$$\lim_{n\to\infty} \mu_n(\psi) = \mu(\psi), \ \psi \in \mathscr{M}$$

Then $\mu_n \rightsquigarrow \mu$.

We note first that, if K is compact, then the sequence $\{\mu_n(K)\}$ is bounded. For let $\psi \in \mathscr{M}$ be such that $\psi \geqslant \chi_K$ (ψ need, for example, only be a sufficiently close approximant to $\phi \in \mathscr{C}_0(\mathbb{R}^p)$ such that $\phi(x) \geqslant 2\chi_K(x)$). Then

$$\mu_n(K) = \int \chi_K(x)\, \mu_n(dx) \leqslant \int \psi(x)\, \mu\,(dx) = \mu_n(\psi);$$

and, since $\{\mu_n(\psi)\}$ is a convergent sequence in $\mathbb{R}$, it is bounded.

Let $\phi \in \mathscr{C}_0(\mathbb{R}^p)$. Then we can find a sequence $\{\psi_m\}$, $\psi_m \in \mathscr{M}$, such that

$$\sup |\, \psi_m(x) - \phi(x)\,| \to 0 \text{ as } m \to \infty$$

and such that there is a compact K with supp $\psi_m \subset K$ and supp $\phi \subset K$. Now

$$|\, \mu_n(\phi) - \mu(\phi)| \leqslant |\, \mu_n(\phi) - \mu_n(\psi_m)\,| + |\, \mu_n(\psi_m) - \psi(\psi_m)\,| + |\, \mu(\psi_m) - \mu(\phi)|$$
$$= T_1 + T_2 + T_3, \text{ say.}$$

Then

$$T_1 \leqslant \int_K |\, \phi(x) - \psi_m(x)\,|\, \mu_n(dx) \leqslant \sup |\, \phi(x) - \psi_m(x)\,|\, \mu_n(K),$$

and this last tends to 0 as $m \to \infty$ uniformly in n. A similar, somewhat simpler argument, shows that $T_3 \to 0$ as $m \to \infty$.

Hence, given $\varepsilon > 0$, we can find m_0 such that

$$T_1 < \tfrac{1}{3}\varepsilon \text{ and } T_3 < \tfrac{1}{3}\varepsilon \text{ for } m \geqslant m.$$

For any fixed $m \geqslant m_0$, $\mu_n(\psi_m) \to \mu(\psi_m)$; and so there is an n_0 such that, for $n \geqslant n_0$, $T_2 < \frac{1}{3}\varepsilon$.

Thus, for $n \geqslant n_0$, $|\, \mu_n(\phi) - \mu(\phi)\,| < \varepsilon$, and the lemma is proved.

LEMMA 1.32. *Suppose that* $\{\lambda_n\}$ and $\{\mu_n\}$ *are sequences of measures in* $\mathbb{R}^p$ *and that* $\lambda_n \rightsquigarrow \lambda$ *and* $\mu_n \rightsquigarrow \mu$. *Then* $\lambda_n \times \mu_n \rightsquigarrow \lambda \times \mu$.

If $\phi \in \mathscr{C}_0(\mathbb{R}^{2p})$ is of the form $f(x)\, g(y)$, $f, g \in \mathscr{C}_0(\mathbb{R}^p)$, then, by Fubini's Theorem,

$$(\lambda_n \times \mu_n)(\phi) = \lambda_n(f)\, \mu_n(g)$$

and this converges to $\lambda(f)\, \mu(g) = (\lambda \times \mu)(\phi)$.

Suppose now that $\phi \in \mathscr{C}_0(\mathbb{R}^{2p})$ is of general form. Choose $r > 0$ so that supp $\phi \subset B_{r-1} \times B_{r-1}$, where B_{r-1} is the open ball in $\mathbb{R}^p$ centred at the origin. Then, by Theorem 1.10, we can find a polynomial p in the variables $x^1, \ldots, x^p; y^1, \ldots, y^p$ such that

$$|\phi(z) - p(z)| < \varepsilon, \quad z \in B_r \times B_r$$

Let

$$\begin{aligned} \hat{p}(x, y) &= p(x, y) & &(x, y) \in B_{r-1} \times B_{r-1} \\ &= p(x, y)\,[r - |x|]\,[r - |y|], & &(x, y) \in (B_r \times B_r) \setminus (B_{r-1} \times B_{r-1}) \\ &= 0, & &(x, y) \in (B_r \times B_r). \end{aligned}$$

Then $\hat{p} \in \mathscr{C}_0(\mathbb{R}^{2p})$ and $\sup_{\mathbb{R}^{2p}} |\phi(z) - \hat{p}(z)| < \varepsilon$.

Thus the set of truncated polynomials, such as $\hat{p}$, is dense in $\mathscr{C}_0(\mathbb{R}^{2p})$. Each $\hat{p}$ is a finite sum of terms of the type $f(x)\, g(y)$, and so

$$(\lambda_n \times \mu_n)(\hat{p}) = (\lambda \times \mu)(\hat{p}),$$

so that, by Lemma 1.31, the result follows.

§ 1.6. Ultraspherical Polynomials

The ultraspherical polynomials (also called Gegenbauer polynomials or, sometimes, Legendre polynomials) $P_{n,p}(\lambda)$, are defined by

$$(1 - 2\rho\lambda + \rho^2)^{(2-p)/2} = \sum_{n=0}^{\infty} P_{n,p}(\lambda)\, \rho^n \tag{1.12}$$

for $p \geqslant 3$ and by

$$-\tfrac{1}{2} \log (1 - 2\rho\lambda + \rho^2) = \sum_{n=1}^{\infty} P_{n,2}(\lambda)\, \rho^n \tag{1.12'}$$

for $p = 2$.†

Setting $\lambda = \cos \alpha$ (so that α is real when $|\lambda| \leqslant 1$ and is purely imaginary when $|\lambda| > 1$), we find that

$$1 - 2\rho\lambda + \rho^2 = (1 - \rho e^{i\alpha})(1 - \rho e^{-i\alpha})$$

† We have $-\log(1 - 2\rho \cos \alpha + \rho^2) = -\log(1 - \rho e^{i\alpha}) - \log(1 - \rho e^{-i\alpha})$

$$= 2 \sum_{n=1}^{\infty} \frac{\rho^n \cos n\alpha}{n}, \text{ so that } P_{n,2}(\cos \alpha) = \frac{1}{n} \cos n\alpha$$

and thus, for $|\lambda| \leqslant 1$, the power series in (12) and (12′) have radius of convergence 1, and for $|\lambda| > 1$ they have radius of convergence $e^{-|\alpha|}$. Within this interval of convergence we may operate freely on the power series.

If we differentiate (1.12) and (1.12′) k times with respect to λ we obtain, for $p \geqslant 2$,

$$(2k+p-4) \ldots (p-2)\, \rho^k\, (1-2\rho\lambda+\rho^2)^{[(2-p)/2]-k} = \sum_{n=1}^{\infty} P_{n,p}^{(k)}(\lambda)\, \rho^n, \tag{1.13}$$

and so $P_{n,p}^{(k)}(\lambda) = 0$ for $k > n$ and $P_{n,p}^{(n)}(\lambda) \neq 0$. Thus **$P_{n,p}(\lambda)$ is a polynomial of degree n precisely.**

If we differentiate (1.12) with respect to ρ, we obtain, for $p \geqslant 3$,

$$(2-p)\,(\rho-\lambda)\,(1-2\rho\lambda+\rho^2)^{-\frac{1}{2}p} = \sum_{n=1}^{\infty} nP_{n,p}(\lambda)\, \rho^{n-1}; \tag{1.14}$$

while from (1.12′) we obtain

$$-(\rho-\lambda)\,(1-2\rho\lambda+\rho^2)^{-1} = \sum_{n=1}^{\infty} nP_{n,2}(\lambda)\, \rho^{n-1}. \tag{1.14′}$$

Also

$$(1-\rho^2)\,(1-2\rho\lambda+\rho^2)^{-\frac{1}{2}p}$$
$$=(1-2\rho\lambda+\rho^2)^{(2-p)/2} - 2\rho(\rho-\lambda)\,(1-2\rho\lambda+\rho^2)^{-\frac{1}{2}p}$$

and, by (1.12) and (1.14), the right hand side becomes, for $p \geqslant 3$,

$$\sum_{n=0}^{\infty} P_{n,p}(\lambda)\, \rho^n \;-\; [2/(2-p)] \sum_{n=0}^{\infty} nP_{n,p}(\lambda)\, \rho^n;$$

and, for $p = 2$, by (1.14′)

$$(1-\rho^2)\,(1-2\rho\lambda+\rho^2)^{-1} = 1+\sum_{n=1}^{\infty} 2nP_{n,2}(\lambda)\, \rho^n$$

so that, for $p \geqslant 3$,

$$(1-\rho^2)\,(1-2\rho\lambda+\rho^2)^{-\frac{1}{2}p} = \sum_{n=0}^{\infty} \frac{2n+p-2}{p-2}\, P_{n,p}(\lambda)\, \rho^n \tag{1.15}$$

and, for $p = 2$,

$$(1-\rho^2)\,(1-2\rho\lambda+\rho^2)^{-1} = 1+\sum_{n=1}^{\infty} 2nP_{n,2}(\lambda)\, \rho^n. \tag{1.15′}$$

It is readily verified that, for $p \geqslant 3$,

$$\frac{\partial}{\partial\rho}\left\{\rho^{p-1} \frac{\partial}{\partial\rho}\,[(1-2\rho\lambda+\rho^2)^{(2-p)/2}]\right\} - \rho^{p-2}\frac{\lambda(p-1)\,(p-2)}{(1-2\rho\lambda+\rho^2)^{\frac{1}{2}p}}$$
$$+\rho^{p-1}\,\frac{\rho(p-2)\,(1-\lambda^2)}{(1-2\rho\lambda+\rho^2)^{\frac{1}{2}p+1}} = 0$$

and, for $p = 2$,

$$\frac{\partial}{\partial\rho}\left\{\rho\,\frac{\partial}{\partial\rho}\left(\tfrac{1}{2}\log\frac{1}{1-2\rho\lambda+\rho^2}\right)\right\}-\frac{\lambda}{1-2\rho\lambda+\rho^2}+\frac{2\rho(1-\lambda^2)}{(1-2\rho\lambda+\rho^2)^2} = 0;$$

and then, using (1.13) with $k = 1$ and $k = 2$, it follows that

$$n(n+p-2)\,P_{n,p}(\lambda)-\lambda(p-1)\,P'_{n,p}(\lambda)+(1-\lambda^2)\,P''_{n,p}(\lambda) = 0. \tag{1.16}$$

Suppose now that $x, a \in \mathbb{R}^p$, that λ is given by

$$|\,x\,|\,|\,a\,|\,\lambda = \sum_{i=1}^{p} x^i a^i$$

and that the Laplacian Δ is defined by

$$\Delta f = \sum_{j=1}^{p} \frac{\partial^2 f}{(\partial x^j)^2}\,.$$

Then

$$\sum_{j=1}^{p}\left(\frac{\partial\lambda}{\partial x^j}\right)^2 = (1-\lambda^2)\,|\,x\,|^{-2}, \sum_{j=1}^{p} x^j\,\frac{\partial\lambda}{\partial x^j} = 0 \text{ and}$$

$$\Delta\lambda = -\lambda(p-1)\,|\,x\,|^{-2},$$

and from this it follows that

$$\Delta\,[|\,x\,|^n\,P_{n,p}(\lambda)] = |\,x\,|^{n-2}\,E(\lambda);\ \Delta\,[|\,x\,|^{2-p-n}\,P_{n,p}(\lambda)]$$
$$= |\,x\,|^{-p-n}\,E(\lambda)$$

where $E(\lambda)$ is the left hand side of (1.16).

Thus $|\,x\,|^n\,P_{n,p}(\lambda)$ and $|\,x\,|^{2-p-n}\,P_{n,p}(\lambda)$ are harmonic functions of x, the first in $\mathbb{R}^p$ and the second in $\mathbb{R}^p \setminus (0)$.

Since, from (1.12) and (1.12′)

$$\sum P_{n,p}(-\lambda)\,(-\rho)^n = \sum P_{n,p}(\lambda)\,\rho^n$$

it follows that $P_{n,p}(\lambda)$ is, for odd n, a polynomial containing only odd powers of λ.

From this it follows that $|\,x\,|^n\,P_{n,p}(\lambda)$ is a homogeneous polynomial in $x^1, \ldots, x^p$. For $P_{n,p}(\lambda)$ is of degree n and, for $k \leqslant n$,

$$|\,x\,|^n\,\lambda^k = |\,x\,|^{n-k}\,(|\,a\,|^{-1}\sum_{i=1}^{p} x^i a^i)^k$$

which, when $(n-k)$ is even, is a homogeneous polynomial in $x^1, \ldots, x^p$ of degree n. Since, for any non-zero term in $P_{n,p}$, $n-k$ is even, the result follows.

Next, for $p \geqslant 3$,

$$|x-a|^{2-p} = (|x|^2-2|x||a|\lambda+|a|^2)^{(2-p)/2}$$

and so, for $|x| < |a|$ and using (1.12)

$$|x-a|^{2-p} = |a|^{2-p}\sum_{n=0}^{\infty} P_{n,p}(\lambda)\frac{|x|^n}{|a|^n}.$$

For $p = 2$,

$$\log\frac{1}{|x-a|} = \log\frac{1}{|a|}-\frac{1}{2}\log\left(1-2\frac{|x|}{|a|}\lambda+\left(\frac{|x|}{|a|}\right)^2\right)$$

and so, for $|x| < |a|$ and using (1.12′),

$$\log\frac{1}{|x-a|} = \log\frac{1}{|a|}+\sum_{n=1}^{\infty} P_{n,2}(\lambda)\frac{|x|^n}{|a|^n}$$

Thus, for $p \geqslant 2$, and for $|x| < |a|$,

$$h(x-a) = h(a)+\sum_{n=1}^{\infty}\frac{|x|^n}{|a|^{n+p-2}}P_{n,p}(\lambda), \tag{1.17}$$

where

$$h(x) = \begin{cases} |x|^{2-p} \text{ for } p \geqslant 3 \\ \log\dfrac{1}{|x|} \text{ for } p = 2 \end{cases}$$

Furthermore, by (1.15), for $|x| < |a|$ and $p \geqslant 3$

$$(|a|^2-|x|^2)|x-a|^{-p} = \sum_{n=o}^{\infty}\frac{2n+p-2}{p-2}\frac{|x|^n}{|a|^{n+p-2}}P_{n,p}(\lambda) \tag{1.18}$$

and by (15′), for $|x| < |a|$ and for $p = 2$,

$$(|a|-|x|^2)|x-a|^{-2} = 1+\sum_{n=1}^{\infty} 2nP_{n,2}(\lambda)\frac{|x|^n}{|a|^n} \tag{1.18′}$$

§ 1.7. Pizzetti's Formula

Suppose that f is a test function and that Δ^r denotes an r-fold application of the Laplacian Δ. Then we have *Pizzetti's Formula*

$$\mathscr{S}_r^a(f) = \sum_{s=0}^{m}(2^s s!\,p(p+2)\ldots(p+2s-2))^{-1}r^{2s}(\Delta^s f)(a)+0(r^{2m+2})$$

We have

$$\mathscr{S}_r^a(f) = \int f(a+t)\, \sigma_r^0(dt)$$

$$= \sum_{n=0}^{2m+1} \frac{1}{n!} \int \left(t_1 \frac{\partial}{\partial a_1} + \ldots + t_p \frac{\partial}{\partial a_p}\right)^n f\, \sigma_r^0(dt) + 0(r^{2m+2}) \qquad (1.19)$$

where the $a_1 \ldots, a_p$ denote that the differential operation on f must be evaluated at a. Now, letting $|k| = k_1 + \ldots + k_p$,

$$\int \left(t_1 \frac{\partial}{\partial a_1} + \ldots + t_p \frac{\partial}{\partial a_p}\right)^n f\, \sigma_r^0(dt)$$

$$= \sum_{|k|=n} \frac{n!}{k_1! \ldots k_p!} \frac{\delta^n f}{\delta a_1^{k_1} \ldots \delta a_p^{k_p}} \left(\int t_1^{k_1} \ldots t_p^{k_p}\, \sigma_r^0(dt)\right). \qquad (1.20)$$

In the integral we substitute spherical polar coordinates to obtain, with $|t| = r$, and writing s_i for $\sin \theta$, and c_i for $\cos \theta_i$,

$$\frac{\Gamma(\frac{1}{2}p)}{2\pi^{\frac{1}{2}p}} r^n \int_0^{2\pi} \int_0^{\pi} \ldots \int_0^{\pi} s_1{}^{k_2+\ldots+k_p+p-2} s_2{}^{k_3+\ldots+k_p+p-3} \quad s_{p-2}^{k_{p-1}+k_p+1} s_{p-1}^{k_p}$$
$$c_1^{k_1} \ldots c_{p-1}^{k_{p-1}} \quad d\theta_1 \ldots d\theta_{p-1}$$

$$= \frac{\Gamma(\frac{1}{2}p)}{2\pi^{\frac{1}{2}p}} r^n \left(\int_0^{\pi} s_1^{l_1} c_1^{k_1} d\theta_1\right) \ldots \left(\int_0^{\pi} s_{p-2}^{l_{p-2}} c_{p-2}^{k_{p-2}} d\theta_{p-2}\right) \left(\int_0^{2\pi} s_{p-1}^{k_p} c_{p-1}^{k_{p-1}} d\theta_{p-1}\right)$$

and, in order that every one of the single integrals (and so the original multiple integral) be non-zero, it is necessary and sufficient that k_1, $k_2 \ldots k_p$ be even. This requires, in particular, that $|k| = n$ be even, and thus in (1.19) only the terms with n even do not vanish.

Now write $2k_1, \ldots 2k_p$ in place of $k_1, \ldots k_p$, $2n$ in place of n. Then the right hand side of (1.20) becomes

$$\frac{\Gamma(\frac{1}{2}p)}{2\pi^{\frac{1}{2}p}} r^{2n} \frac{\partial^{2n} f}{\partial a_1^{2k_1} \ldots \partial a_p^{2k_p}} \frac{\Gamma(k_1+\frac{1}{2}) \ldots \Gamma(k_p+\frac{1}{2})}{\Gamma(n+\frac{1}{2}p)} \frac{(2n)!}{(2k_1)! \ldots (2k_p)!}$$

which, after simplification, becomes

$$\frac{(2n)!}{n!} \frac{r^{2n}}{2^{2n}} \frac{\Gamma(\frac{1}{2}p)}{\Gamma(n+\frac{1}{2}p)} \frac{n!}{k_1! \ldots k_p!} \left(\frac{\partial^2}{\partial a^2}\right)^{k_1} \cdots \left(\frac{\partial^2}{\partial a^2}\right)^{k_p} f$$

so that the $2n$-th term in the right hand side of (19) becomes

$$\frac{1}{n!} \frac{r^{2n}}{2^{2n}} \frac{\Gamma(\frac{1}{2}p)}{\Gamma(n+\frac{1}{2}p)} \left(\frac{\partial^2}{\partial a_1^2} + \cdots + \frac{\partial^2}{\partial a_p^2}\right)^n f,$$

and Pizzetti's Formula now follows.

Clearly, to ask that f be a test function is to ask more than is necessary. It is enough to suppose that f is $(2m+2)$-times continuously differentiable.

From Pizzetti's Formula we pass easily to the corresponding result for ball means. We have

$$r^p \mathscr{B}_r^a(f) = p\int \rho^{p-1} \mathscr{S}_\rho^a(f)\, d\rho$$

$$= \sum_{s=0}^{m} (2^s s!\, p(p+2) \dots (p+2s-2))^{-1} (\Delta^s f)(a)\, p \int_0^r \rho^{2s+p-1}\, d\rho$$

$$+ O(r^{2m+p+2})$$

$$= \sum_{s=0}^{m} (2^s s!\, (p+2) \dots (p+2s))^{-1} r^{2s} (\Delta^s f)(a) + O(r^{2m+2})$$

whence

$$\mathscr{B}_r^a(f) = \sum_{s=0}^{m} (2^s\, s!(p+2) \dots (p+2s))^{-1} r^{2s} (\Delta^s f)(a) + O(r^{2m+2}).$$

Chapter 2

SUPERHARMONIC, SUBHARMONIC AND HARMONIC FUNCTIONS IN $\mathbb{R}^p$

We depart somewhat from the usual treatment of this topic by compactifying $\mathbb{R}^p$. Let $\overline{\mathbb{R}}^p$ denote the topological space obtained by adjoining to $\mathbb{R}^p$ the Alexandroff point ω as follows:

Let

$$\overline{\mathbb{R}}^p = \mathbb{R}^p \cup (\omega)$$

and say that any set $G \cup (\omega)$, where $G \subset \mathbb{R}^p$ is open and contains a set of the form $\mathbb{R}^p \setminus B_r(a)$, is a neighbourhood of ω. This amounts to saying that the open neighbourhoods of ω are the complements of compact sets in $\mathbb{R}^p$.†

Thus while $\mathbb{R}^p$ is a locally compact topological space $\overline{\mathbb{R}}^p$ is a compact topological space and it is from the compactness of $\overline{\mathbb{R}}^p$ that many simplifications result.

§ 2.1. Superharmonic and Subharmonic Functions

Let $G \subset \overline{\mathbb{R}}^p$ be open. A numerically-valued function f defined in G is said to be superharmonic in the wide sense, or **hyperharmonic,** in G if it has the following properties:

(i) f is l.s.c. in G,

(ii) $f(x) > -\infty$ for all $x \in G$,

(iii) If $a \in G \setminus (\omega)$ and $B_r(a) \subset G$, then

$$f(a) \geqslant \mathscr{S}_r^a(f)$$

and, if $\omega \in G$,

$$f(\omega) \geqslant \mathscr{S}_r^b(f)$$

whenever $b \in \mathbb{R}^p$ and $\overline{\mathbb{R}}^p \setminus B_r(b) \subset G$.

† Note that $\overline{\mathbb{R}}^p$ for $p \geqslant 2$ is gained from $\mathbb{R}^p$ by adding only one ideal element, whereas $\overline{\mathbb{R}}$ has two ideal elements. Thus $\overline{\mathbb{R}}^p$ is akin to the Neumann sphere in complex analysis rather than to $\overline{\mathbb{R}}$.

Some important properties of hyperharmonic functions follow immediately from the definition.

THEOREM 2.1. (*a*) *Suppose that* $f_1, \ldots, f_k$ *are hyperharmonic in* G *and that* $\lambda_1, \ldots, \lambda_k$ *are non-negative constants. Then* $\sum_{r=1}^{k} \lambda_r f_r$ *is hyperharmonic in* G.

(*b*) *Suppose that* $f_1, \ldots, f_k$ *are hyperharmonic in* G. *Then*

$$g(x) = \inf_{1 \leqslant r \leqslant k} (f_r(x)) \text{ is hyperharmonic in } G.$$

(*c*) *Suppose that* $\{f_\alpha\}_{\alpha \in A}$ *is an up-directed family and that each* f_α *is hyperharmonic in* G. *Then* $g(x) = \sup_A f_\alpha(x)$ *is hyperharmonic in* G.

First, by Theorem 1.1, $g(x) = \sum_{r=1}^{k} \lambda_r f_r(x)$ is l.s.c. in G, since $f_1, \ldots, f_k$ are. Next, $g(x) > -\infty$ for all $x \in G$, since $f_r(x) > -\infty$ $(r = 1, 2, \ldots, k)$. Finally, for $a \in \mathbb{R}^p$,

$$g(a) = \sum_{r=1}^{k} \lambda_r f_r(a) \geqslant \sum_{r=1}^{k} \lambda_r \mathscr{S}_\rho^a(f_k) = \mathscr{S}_\rho^a(g)$$

for all appropriate ρ, and a similar argument shows that $g(\omega) \geqslant \mathscr{S}_\rho^b(g)$. This gives (*a*).

By Theorem 1.1, g is l.s.c. in G and, next, $g(x) > -\infty$, for all $x \in G$. Suppose that

$$a \in G \setminus (\omega) \text{ and that } f_s(a) = \inf_{1 \leqslant r \leqslant k} f_r(a).$$

Then, for suitable ρ,

$$g(a) = f_s(a) \geqslant \mathscr{S}_\rho^a(f_s) \geqslant \mathscr{S}_\rho^a(g)$$

since $f_s(x) \geqslant g(x)$ for all $x \in S_\rho(a)$. Thus $g(a) \geqslant \mathscr{S}_\rho^a(g)$, and a similar argument gives the required property for ω, if $\omega \in G$.

We define h, the **fundamental function**, by

$$h(x) = \begin{cases} \log 1/|x| & \text{when } p = 2 \\ |x|^{2-p} & \text{when } p \geqslant 3 \end{cases}$$

and we shall show later that h is superharmonic in $\mathbb{R}^p$ (although not in $\overline{\mathbb{R}}^p$!).

Assuming this for the moment, we may note that (*b*) does not extend to the case when the infimum is taken over an infinite family. Thus, if

$f_r(x) = \frac{1}{r} h(x)$ $(r = 1, 2, \ldots)$ and $G = B_1(0)$, then

$$g(x) = \inf_r f_r(x) = +\infty \quad \text{when } x = 0$$
$$= 0 \quad \text{when } x \neq 0,$$

and then g is not l.s.c. in G and thus not hyperharmonic.

We turn to (*c*). By Theorem 1.6, g is l.s.c. in G. Next, $g(x) > -\infty$ for all $x \in G$ and, finally, for $a \in G \setminus (\omega)$,

$$g(a) \geqslant f_\alpha(a) \geqslant \mathscr{S}_\rho^a(f_\alpha) \quad \text{for } \alpha \in A.$$

Now, by Theorem 1.18,

$$\sup_A \mathscr{S}_\rho^a(f_\alpha) = \mathscr{S}_\rho^a(g)$$

and hence $g(a) \geqslant \mathscr{S}_\rho^a(g)$. A similar argument suffices to deal with the case when $\omega \in G$.

There is not a similar result for a down-directed family of hyperharmonic functions. The counter-example following (*b*) is also a down-directed family, of which the infimum is not hyperharmonic. But we shall see later (Theorem 4.35) that, in fact, the infimum of any family of hyperharmonic functions which is locally bounded below, is, in a sense to be made precise, nearly a hyperharmonic function.

We may note here that, given G, the function f given by $f(x) = +\infty$ for all $x \in G$ is hyperharmonic in G. The following theorem shows that this function is, in some sense, a crucial example.

THEOREM 2.2. *Suppose that f is hyperharmonic in the domain $G \subset \mathbb{R}^p$. Then either $f(x) = +\infty$ everywhere in G, or f is locally integrable in G.*

First, if f is hyperharmonic in G, then for $a \in G$ and $B_r(a) \subset G$

$$f(a) \geqslant \mathscr{B}_r^a(f) \tag{2.1}$$

For, by (1.9),

$$\mathscr{B}_r^a(f) = pr^{-p} \int_0^r \rho^{p-1} \mathscr{S}_\rho^a(f)\, d\rho$$

and this last is not greater than

$$pr^{-p} f(a) \int_0^r \rho^{p-1}\, d\rho = f(a)$$

In parenthesis, we have shown that if f is hyperharmonic then (2.1) holds. In fact, as we shall see later, (2.1) is a **characterisation** for hyperharmonic functions, together with the semi-continuity requirement. Let

$$S = \{x \in G \mid f \text{ is integrable over some neighbourhood of } x\}$$

Let $x \in S$. Then there is an $r > 0$ such that f is integrable over $B_r(x)$ and, consequently, $|f|$ is so integrable. Let $y \in B_{\frac{1}{2}r}(x)$. Then $B_{\frac{1}{2}r}(y) \subset B_r(x)$ and so $|f|$ is integrable over $B_{\frac{1}{2}r}(y)$. Since, then, f is integrable over $B_{\frac{1}{2}r}(y)$, it follows that $y \in S$. Thus, when $x \in S$, $B_{\frac{1}{2}r}(x) \subset S$ and so S is open.

Next, let $x \in G \setminus S$. Then f is not integrable in any neighbourhood of x. Also, since f is l.s.c., it is bounded below in any bounded neighbourhood of x and so $\mathscr{B}_r^x(f) = +\infty$ whenever $B_r(x) \subset G$. From (2.1) it then follows that $f(x) = +\infty$.

Suppose $B_r(x) \subset G$ and that $y \in B_{\frac{1}{3}r}(x)$. Then $B_{\frac{2}{3}r}(y) \supset B_{\frac{1}{3}r}(x)$ and so f is not integrable in $B_{\frac{2}{3}r}(y)$. Consequently $f(y) = +\infty$. Thus f is not integrable in any neighbourhood of y so that $y \in G \setminus S$. Thus, when $x \in G \setminus S$, $B_{\frac{1}{3}r}(x) \subset G \setminus S$ and so $G \setminus S$ is open.

Since $G = S \cup (G \setminus S)$ and G is a domain, and so connected, we must have either $S = \phi$ or $G \setminus S = \phi$, which gives Theorem 2.1.

There is thus a (nearly) all-or-none principle at work here, and we say that f is **superharmonic** in the domain G if it is hyperharmonic but not identically infinite there. We note that a superharmonic function is finite a.e.

THEOREM 2.3. *If f is superharmonic in the domain $G \subset \overline{\mathbb{R}}^p$ and attains its infimum in G then it is constant in G.*

Let $\lambda = \inf_G f$, and suppose that it is attained in G. It follows that λ is finite.

Let $S = \{x \in G \mid f(x) = \lambda\}$ and suppose that $x \in S \setminus (\omega)$. Then, by (2.1) $\lambda \geqslant \mathscr{B}_r^x(f)$. But since $f(y) \geqslant \lambda$ we have also that $\lambda \leqslant \mathscr{B}_r^x(f)$, and hence $\lambda = \mathscr{B}_r^x(f)$ for all suitable r. Thus

$$\lambda m B_r(x) = \int_{B_r(x)} f(y)\, dy.$$

We assert now that $S \setminus (\omega)$ is open. For suppose that there is no neighbourhood of x in which $f(y) = \lambda$. Then, given $r > 0$, there is $y \in B_r(x)$ and $k > 0$ such that $f(y) = \lambda + 2k$. Since f is l.s.c., there is $\rho > 0$ such that $f(z) > \lambda + k$ whenever $z \in B_\rho(y)$, and we may suppose $B_\rho(y) \subset B_r(x)$. Thus

$$\int_{B_r(x)} f(y)\, dy = \int_{B_\rho(y)} f(z)\, dz + \int_{B_r(x) \setminus B_\rho(y)} f(z)\, dz \geqslant (\lambda + k) m B_\rho(y) + \lambda m(B_r(x) \setminus B_\rho(y)),$$

and this last is greater than $\lambda m B_r(x)$—a contradiction. Thus there is a neighbourhood of x in which $f(y) = \lambda$ and so $S \setminus (\omega)$ is open.

If $\omega \in S$, then $f(\omega) = \lambda$ and so, for $\overline{\mathbb{R}}^p \setminus B_r(x) \subset G$,

$$\lambda \geqslant \mathscr{S}_r^x(f).$$

Consequently, for $R > r$,

$$\lambda \int_r^R \rho^{p-1}\, d\rho \geqslant \int_r^R \rho^{p-1} \mathscr{S}_\rho^x(f)\, d\rho$$

whence, by (1.10), $\lambda \geqslant \mathscr{B}_{r,R}^x(f)$. Also, since λ is the infimum, $\lambda \leqslant \mathscr{B}_{r,R}^x(f)$ and so

$$\lambda(m B_R(x) - m B_r(x)) = \int_{B_R(x) \setminus B_r(x)} f(y)\, dy$$

Suppose, if possible, that there is no neighbourhood of ω in which $f(y) = \lambda$. Then there is an $r > 0$ and $y \in \mathbb{R}^p \setminus \bar{B}_r(x)$ and $k > 0$ such that $f(y) = \lambda + 2k$. Also there is a $\rho > 0$ such that $f(z) > \lambda + k$ whenever $z \in B_\rho(y)$ and we may suppose $B_\rho(y) \subset \mathbb{R}^p \setminus \bar{B}_r(x)$. Choose R so that $B_\rho(y) \subset B_R(x)$. Then

$$\int_{B_R(x) \setminus B_r(x)} f(y)\, dy = \int_{B_\rho(y)} f(z)\, dz + \int_{B_R(x) \setminus (B_r(x) \cup B_\rho(y))} f(z)\, dz$$

and this is not less than $(\lambda+k)mB_\rho(y) + \lambda m\ [B_R(x) \setminus (B_r(x) \cup B_\rho(y))]$ which is greater than $\lambda m\ (B_R(x) \setminus B_r(x))$; which gives a contradiction. Thus S is open.

Again $G \setminus S = \{x \in G \mid f(x) > \lambda\}$ and this, by Theorem 1.3, is open. Since G is connected, either S or $G \setminus S$ is void; and this gives Theorem 2.3.

COROLLARY 2.3.1. *If f is superharmonic in $\overline{\mathbb{R}}^p$ it is constant.*

Since f is l.s.c. in $\overline{\mathbb{R}}^p$, it attains its finite infimum in the compact set $\overline{\mathbb{R}}^p$. It is therefore constant in the domain $\mathbb{R}^p$.

From Theorem 2.3, one is tempted to deduce, in a rough and ready way, that a function superharmonic in a set G takes its minimum on the boundary of G. But since, strictly speaking, f need not be defined on ∂G, the precise theorem must take a more sophisticated form. However, if f is defined in an open set containing $\bar{G}$, then the following theorem would imply the naive form indicated above.

THEOREM 2.4. *If f is superharmonic in the domain $G \subset \mathbb{R}^p$, then, provided $\partial G \neq \phi$,*

$$\inf_{x \in G} f(x) = \inf_{\xi \in \partial G} [\liminf_{x \in G,\, x \to \xi} f(x)] \tag{2.2}$$

Define the function F in $\bar{G}$ by

$$\begin{aligned} F(x) &= f(x),\ x \in G \\ &= \liminf_{y \in G,\, y \to x} f(y),\ x \in \partial G \end{aligned}$$

and let $\lambda = \inf_G F$ and μ be the right hand side of (2). Suppose, if possible, that $\lambda < \mu$. Then there is $\nu < \mu$ such that $\{x \in \bar{G} \mid F(x) \leqslant \nu\} = \{x \in G \mid f(x) \leqslant \nu\}$ is non-void. Since F is l.s.c. in G, by Theorem 1.3, this set is closed and so compact. Thus f attains its infimum in this set, and this infimum is λ. Consequently f attains its infimum in G and so is constant; but then $\mu = \lambda$. Consequently, in any event, $\lambda \geqslant \mu$. But clearly $\lambda \leqslant \mu$ and so $\lambda = \mu$. This gives the theorem.

In the more usual account of the theory (one not utilising the Alexandroff point), Corollary 2.3.1 has no counterpart, since, for example, $h(x)$ is superharmonic in $\mathbb{R}^p$ and is not constant. Theorem 2.4 is true only when, in addition, the domain G is bounded as the example $h(x)$ in $G = \mathbb{R}^p \setminus B_1(0)$ shows. In this account we get neater theorems—but at a price—we exact the condition that f is superharmonic at ω. The counter-example we have just given is not a counter-example to Theorem 2.4 precisely because either ω is a frontier point, at which the minimum is attained, or because $\omega \in G$ and then f is not superharmonic at ω.

A function f defined in an open subset G or $\overline{\mathbb{R}}^p$ is said to be subharmonic in the wide sense (or **hypoharmonic**) in G if it has the following properties:

(i) $f(x) < +\infty$ for all $x \in G$,

(ii) f is u.s.c. in G,

(iii) If $a \in G \setminus (\omega)$ and $B_r(a) \subset G$, then

$$f(a) \leqslant \mathscr{S}_r^a(f)$$

and, if $\omega \in G$,

$$f(\omega) \leqslant \mathscr{S}_r^b(f) \text{ whenever } \overline{\mathbb{R}}^p \setminus B_r(b) \subset G.$$

It follows immediately that f is hypoharmonic in G if, and only if, $-f$ is hyperharmonic in G. In consequence, we have

THEOREM 2.5. (*a*) *Suppose that $f_1, \ldots, f_k$ are hypoharmonic in G and that $\lambda_1, \ldots, \lambda_k$ are non-negative constants. Then $\sum_{r=1}^{k} \lambda_r f_r$ is hypoharmonic in G.*

(*b*) *Suppose that $f_1, \ldots, f_k$ are hypoharmonic in G. Then $g(x) = \sup_{1 \leqslant r \leqslant k} f_r(x)$ is hypoharmonic in G. This result does not hold for infinite families.*

(*c*) *Suppose that $\{f_\alpha\}_{\alpha \in A}$ is a down-directed family and that each f_α is hypoharmonic in G. Then $g(x) = \inf_A f_\alpha(x)$ is hypoharmonic in G.*

THEOREM 2.6. *Suppose that f is hypoharmonic in the domain G. Then either $f(x) = -\infty$ everywhere in G or f is locally integrable in G and hence f is finite a.e. in G.*

If the second alternative holds, we say that f is **subharmonic** in G.

THEOREM 2.7. *If f is subharmonic in the domain G and attains its supremum in G, then it is constant in G.*

COROLLARY 2.7.1. *If f is subharmonic in $\overline{\mathbb{R}}^p$, it is constant.*

THEOREM 2.8. *If f is subharmonic in the domain G, then, provided $\partial G \neq \phi$,*

$$\sup_{x \in G} f(x) = \sup_{\xi \in \partial G} [\limsup_{x \in G, x \to \xi} f(x)].$$

A function which is both superharmonic and subharmonic in an open set G (note that a function cannot be both hyperharmonic and hypoharmonic if it is not superharmonic and subharmonic) is said to be **harmonic in G.** Thus, f is harmonic in G if, and only if,

(i) f is continuous and real-valued in G

(ii) If $a \in G \setminus (\omega)$ and $B_r(a) \subset G$ then

$$f(a) = \mathcal{S}_r^a(f)$$

and, if $\omega \in G$,

$$f(\omega) = \mathcal{S}_r^b(f) \text{ whenever } \mathbb{R}^p \setminus B_r(a) \subset G.$$

It follows immediately that

(a) If $f_1, \ldots, f_k$ are harmonic in G and $\lambda_1, \ldots \lambda_k$ are real constants then $\sum_{r=1}^{k} \lambda_r f_r$ is harmonic in G.

(b) Suppose that $f_1, \ldots, f_k$ are harmonic in G. Then $\sup_{1 \leqslant r \leqslant k} f_r(x)$ is subharmonic in G and $\inf_{1 \leqslant r \leqslant k} f_r(x)$ is superharmonic in G.

(c) The supremum of an up-directed family of harmonic functions is hyperharmonic and the infimum of a down-directed family of harmonic functions is hypoharmonic.

(*b*) cannot be improved since the sup and inf there need not be harmonic. On the other hand, as we shall see later (Theorems 2.24 and 2.25), (*c*) can be greatly improved.

It also follows readily that

THEOREM 2.9. *If f is harmonic in a domain G and attains either its infimum or supremum in G, then it is constant in G.*

COROLLARY 2.9.1. *If f is harmonic in $\mathbb{R}^p$, it is constant.*

THEOREM 2.10. *If f is harmonic in the domain G, then, provided $\partial G \neq \phi$*

$$\inf_{x \in G} f(x) = \inf_{\xi \in \partial G} [\liminf_{x \in G, x \to \xi} f(x)]; \quad \sup_{x \in G} f(x) = \sup_{\xi \in \partial G} [\limsup_{x \in G, x \to \xi} f(x)]$$

The next theorem states a result which makes clear why the terms 'superharmonic' and 'subharmonic' are used.

THEOREM 2.11. *Let G be a domain in $\mathbb{R}^p$. Then*

(*i*) *If f is superharmonic in G, u is harmonic in G, and*

$$\inf_{\xi \in \partial G} [\liminf_{x \in G, x \to \xi} (f(x) - u(x))] \geqslant 0,$$

then either $f(x) > u(x)$ in G or $f(x) = u(x)$ everywhere in G.

(*ii*) *If f is subharmonic in G, u is harmonic in G, and*

$$\sup_{\xi \in \partial G} [\limsup_{x \in G, x \to \xi} (f(x)-u(x))] \leqslant 0,$$

then either $f(x) < u(x)$ in G or $f(x) = u(x)$ everywhere in G.

First, $f(x)-u(x)$ is superharmonic in G and so, by Theorem 2.4, $\inf_{x \in G} [f(x)-u(x)] \geqslant 0$. Then either $f(x) > u(x)$ in G or $f(x)-u(x)$ attains its infimum in G; in which case, by Theorem 2.3, $f(x) = u(x)$ everywhere. This proves (*i*).

(*ii*) follows similarly by using Theorems 2.7 and 2.8 or, alternatively, by noting that $-f$ is superharmonic.

§ 2.2. Laplacians

Suppose that f is defined in an open set $G \subset \mathbb{R}^p$, and has continuous second derivatives there. Then we define the Laplacian Δf of f by

$$\Delta f = \sum_{r=1}^{p} \frac{\partial^2 f}{\partial x_r^2}.$$

It can be shown, under the transformation (1.6), that

$$\rho^{p-1} \sin^{p-2}\theta_1 \sin^{p-3}\theta_2 \ldots \sin\theta_{p-2} (\Delta f)(x)$$

$$= \frac{\partial}{\partial \rho}\left(\rho^{p-1} \sin^{p-2}\theta_1 \ldots \sin\theta_{p-2} \frac{\partial f}{\partial \rho}\right)$$

$$+ \frac{\partial}{\partial \theta_1}\left(\rho^{p-3} \sin^{p-2}\theta_1 \ldots \sin\theta_{p-2} \frac{\partial f}{\partial \theta_1}\right)$$

$$+ \sum_{r=2}^{p-2} \frac{\partial}{\partial \theta_r}\left(\rho^{p-3} \sin^{p-4}\theta_1 \ldots \sin^{p-r-2}\theta_{r-1} \sin^{p-r-1}\theta_r \ldots \sin\theta_{p-2} \frac{\partial f}{\partial \theta_r}\right)$$

$$+ \frac{\partial}{\partial \theta_{p-1}}\left(\rho^{p-3} \sin^{p-4}\theta_1 \ldots \sin\theta_{p-4} (\sin\theta_{p-2})^{-1} \frac{\partial f}{\partial \theta_{p-1}}\right),$$

and from this form of the Laplacian we may deduce a useful particular case of Green's Formula.

LEMMA 2.12. *If f and g have continuous second derivatives in $G \subset \mathbb{R}^p$, then, when $A = \bar{B}_R(a) - B_r(a) \subset G$ with $R > r \geqslant 0$ and with $\rho = |x-a|$ we have*

$$\int_A (f\Delta g - g\Delta f)dx = \frac{2\pi^{\frac{1}{2}p}}{\Gamma(\frac{1}{2}p)}\left\{R^{p-1}\int\left(f\frac{\partial g}{\partial \rho} - g\frac{\partial f}{\partial \rho}\right)\sigma_R^a(dx)\right.$$

$$\left. - r^{p-1}\int\left(f\frac{\partial g}{\partial \rho} - g\frac{\partial f}{\partial \rho}\right)\sigma_r^a(dx)\right\}.$$

Transforming to spherical polar coordinates with origin at a and denoting $\sin\theta_r$ by s_r, we have

$$\int_A (f\Delta g - g\Delta f)dx$$

$$= \int_0^{2\pi}\int_0^{\pi}\cdots\int_0^{\pi} s_1^{p-2}\ldots s_{p-2}\left\{\int_r^R f\frac{\partial}{\partial\rho}\left(\rho^{p-1}\frac{\partial g}{\partial\rho}\right)\right.$$

$$\left.-g\frac{\partial}{\partial\rho}\left(\rho^{p-1}\frac{\partial f}{\partial\rho}\right)d\rho\right\}d\theta_1\ldots d\theta_{p-1} + \int_r^R\int_0^{2\pi}\int_0^{\pi}\cdots\int_0^{\pi}\rho^{p-3}s_2^{p-3}\ldots s_{p-2}$$

$$\left\{\int_0^{\pi}\left[f\frac{\partial}{\partial\theta_1}\left(s_1^{p-2}\frac{\partial g}{\partial\theta_1}\right)-g\frac{\partial}{\partial\theta_1}\left(s_1^{p-2}\frac{\partial f}{\partial\theta_1}\right)\right]d\theta_1\right\}d\theta_2\ldots d\theta_{p-1}\,d\rho$$

$$+\sum_{r=2}^{p-2}\int_r^R\int_0^{2\pi}\int_0^{\pi}\cdots\int_0^{\pi}\rho^{p-3}s_1^{p-4}\ldots s_{r-1}^{p-r-2}s_{r+1}^{p-r-2}\ldots s_{p-2}$$

$$\left\{\int_0^{\pi}\left[f\frac{\partial}{\partial\theta_r}\left(s_r^{p-r-1}\frac{\partial g}{\partial\theta_r}\right)-g\frac{\partial}{\partial\theta_r}\left(s_r^{p-r-1}\frac{\partial f}{\partial\theta_r}\right)\right]d\theta_r\right\}d\theta_1\ldots d\theta_{p-2}\,d\rho$$

$$+\int_r^R\int_0^{\pi}\cdots\int_0^{\pi}\rho^{p-3}s_1^{p-4}\ldots s_{p-4}s_{p-2}^{-1}\left\{\int_0^{2\pi}\left(f\frac{\partial^2 g}{\partial\theta_{p-1}^2}-g\frac{\partial^2 f}{\partial\theta_{p-1}^2}\right)d\theta_{p-1}\right\}$$

$$d\theta_{p-2}\ldots d\theta_1\,d\rho.$$

In each term on the right we integrate the inmost integral by parts and find that the second to the $(p-1)$-st terms vanish for obvious reasons, and the p-th term does so because f, g, $\dfrac{\partial f}{\partial\theta_{p-1}}$ and $\dfrac{\partial g}{\partial\theta_{p-1}}$ have period 2π in θ_{p-1}. The inmost integral in the first term becomes

$$\rho^{p-1}\left(f\frac{\partial g}{\partial\rho}-g\frac{\partial f}{\partial\rho}\right)\Bigg]_{\rho=r}^{\rho=R},$$

and so the term itself written in terms of spherical measures becomes

$$\frac{2\pi^{\frac{1}{2}p}}{\Gamma(\frac{1}{2}p)}\left\{R^{p-1}\int\left(f\frac{\partial g}{\partial\rho}-g\frac{\partial f}{\partial\rho}\right)\sigma_R^a(dx)-r^{p-1}\int\left(f\frac{\partial g}{\partial\rho}-g\frac{\partial f}{\partial\rho}\right)\sigma_r^a(dx)\right\},$$

which is the required result.

We now show that the definition that has been given of harmonicity is tantamount to the more usual one, that of the vanishing of the Laplacian.

THEOREM 2.13. *If f is continuously twice differentiable in $G \subset \mathbb{R}^p$, then f is harmonic in G if, and only if,*

$$\Delta f = 0$$

in G.

Using Lemma 2.12 with $r = 0$ and $g = 1$, we have, when $\bar{B}_R(a) \subset G$,

$$\int_{B_R(a)} \Delta f \, dx = R^{p-1} \frac{2\pi^{\frac{1}{2}p}}{\Gamma(\frac{1}{2}p)} \left\{ \int \frac{\partial f}{\partial \rho} \, \sigma_R^a(dx) \right\}$$

The integral in brackets on the right may be rewritten as

$$\frac{d}{dR} \{ \int f(x) \, \sigma_R^a(dx) \}.$$

If f is harmonic in G, this last integral has the value $f(a)$ and so does not vary with R. Hence $\int_{B_R(a)} \Delta f dx = 0$ and since this holds for all $B_R(a) \subset G$ it follows that Δf, being continuous, has the value 0 in G.

Conversely, if $\Delta f = 0$ in G, then $\int f(x) \, \sigma_R^a(dx)$ is independent of R. Letting $R \to 0$, since f is continuous, it follows that this constant value is $f(a)$ and so f is harmonic in G.

In Theorem 2.13 the requirement that f be continuously twice differentiable turns out to be redundant, since, as we shall see later, a harmonic function is infinitely differentiable (and indeed analytic).

LEMMA 2.14. *Suppose that f is twice continuously differentiable in G and that $\bar{B}_R(a) \subset G$. Then, if $h_a(x) = h(x-a)$, we have, for $p \geqslant 3$,*

$$f(a) = \int f(x) \, \sigma_R^a(dx) + \frac{R}{p-2} \int \frac{\partial f}{\partial \rho} \, \sigma_R^a(dx) - \frac{\Gamma(\frac{1}{2}p)}{(p-2) \, 2\pi^{\frac{1}{2}p}} \int_{B_R(a)} h_a(x) \, \Delta f \, dx.$$

When $p = 2$, we have a similar formula in which the $R/(p-2)$ in the second term is replaced by $R \log R$ and the $p-2$ in the third term by unity.

Setting $g = h_a$, we have $\dfrac{\partial g}{\partial \rho} = (2-p)\rho^{1-p}$ when $p \geqslant 3$, and we replace $2-p$ by -1 when $p = 2$. Also $\Delta g = 0$ when $x \neq a$. Using Lemma 2.12 with this g and with $p \geqslant 3$, we have

$$\frac{\Gamma(\frac{1}{2}p)}{2\pi^{\frac{1}{2}p}} \int_A h_a \Delta f dx = R \int \frac{\partial f}{\partial \rho} \, \sigma_R^a(dx) + (p-2) \int f(x) \, \sigma_R^a(dx) - r \int \frac{\partial f}{\partial \rho} \, \sigma_R^a(dx) - (p-2) \int f(x) \, \sigma_r^a(dx).$$

Letting $r \to 0$, we then have the required result. A similar argument gives the result for $p = 2$.

In particular, when f is continuously twice differentiable in $\mathbb{R}^p$ and has compact support, we have, choosing R so that $\operatorname{supp} f \subset B_R(a)$, that

$$f(a) = -\frac{\Gamma(\tfrac{1}{2}p)}{(p-2)\,2\pi^{\frac{1}{2}p}} \int_{B_R(a)} h_a(x)\,(\Delta f)(x)\,dx \tag{2.3}$$

when $p \geqslant 3$, and $(p-2)$ is replaced by 1 when $p = 2$.

§ 2.3. The Poisson Integral

First, some preliminary lemmas.

LEMMA 2.15. *Let $p \geqslant 2$. Then*

$$\int h(x-t)\,\sigma_R^0(dt) = \begin{cases} h(R) & \text{for } |x| \leqslant R \\ h(x) & \text{for } |x| > R \end{cases}$$

where, by an abuse of notation, we denote $h(x)$ by $h(R)$ when $|x| = R$.

This is a particular case of Lemma 3.22.

LEMMA 2.16. *Let $p \geqslant 2$. Then*

$$\int \frac{R^2-|x|^2}{|x-t|^p}\,\sigma_R^0(dt) = \begin{cases} R^{2-p} & \text{for } |x| < R \\ -|x|^{2-p} & \text{for } |x| > R. \end{cases}$$

We have

$$\frac{R^2-r^2}{(R^2-2Rr\cos\theta+r^2)^{\frac{1}{2}p}} = \frac{2R\,(R-r\cos\theta)}{(R^2-2Rr\cos\theta+r^2)^{\frac{1}{2}p}} - \frac{1}{(R^2-2Rr\cos\theta+r^2)^{\frac{1}{2}p-1}}$$

When $r < R$ and $p \geqslant 3$, choosing coordinates so that $x = (r, 0, \ldots, 0)$, Lemma 2.15 gives

$$\int (R^2-2Rr\cos\theta_1+r^2)^{1-\frac{1}{2}p}\,\sigma_R^0(dt) = R^{2-p} \quad \text{for } |x| < R$$

and, differentiating with respect to R, we have

$$\int (R-r\cos\theta_1)\,(R^2-2Rr\cos\theta_1+r^2)^{-\frac{1}{2}p}\,\sigma_R^0(dt) = R^{1-p}.$$

Consequently,

$$\int (R^2-r^2)\,(R^2-2Rr\cos\theta_1+r^2)^{-\frac{1}{2}p}\,\sigma_R^0(dt) = 2R^{2-p}-R^{2-p} = R^{2-p}.$$

When $r > R$ we interchange the above roles of r and R to obtain the second result. A similar argument gives the result for $p = 2$.

We have already mentioned the fact that $h(x)$ is harmonic in $\mathbb{R}^p\setminus(0)$, and we note that this can be deduced from Lemma 2.15, since $h(x)$ is clearly continuous in $\mathbb{R}^p \setminus (0)$. Furthermore, setting $x = 0$ in Lemma 2.15, we see that $\mathscr{S}_R^0(h) = h(R) < +\infty$ and so h is superharmonic at 0, since it is l.s.c. there. Thus h is superharmonic in $\mathbb{R}^p$ (but not, as we have seen, in $\overline{\mathbb{R}}^p$).

We could also prove the harmonicity of $h(x)$ by using the polar form of the Laplacian. Also, choosing coordinates suitably,

$$\frac{R^2 - |x|^2}{|x-t|^p} = \frac{2R \cos \theta_1}{\rho^{p-1}} - \frac{1}{\rho^{p-2}}$$

the polar form of the Laplacian shows that this function (as a function of x) is harmonic in $\mathbb{R}^p \setminus (t)$.

We may now turn to the main concern of this section. The **Poisson kernel** $K_R^a(x, t)$ is defined, for $x \notin S_R(a)$, by

$$K_R^a(x, t) = \operatorname{sgn}(R - |x-a|)\, R^{p-2} (R^2 - |x-a|^2)\, |x-t|^{-p}.$$

It is a harmonic function of x in $\mathbb{R}^p \setminus (S_R(a) \cup (t))$.

Given a function f defined and continuous on $S_R(a)$, we define its **Poisson integral** $I_R^a f$ by

$$(I_R^a f)(x) = \int f(t)\, K_R^a(x, t)\, \sigma_R^a(dt)$$

Then $I_R^a f$ is harmonic in $\mathbb{R}^p \setminus S_R(a)$ and we have

THEOREM 2.17. *Let f be defined and continuous in $S_R(a)$. Then*

$$\lim_{x \to \xi} (I_R^a f)(x) = f(\xi) \quad \text{for } \xi \in S_R(a) \tag{2.4}$$

where x tends to ξ without entering $S_R(a)$.

Suppose first that $x \in B_R(a)$. Then we have, by Lemma 2.16, that

$$(I_R^a f)(x) - f(\xi) = \int (f(t) - f(\xi))\, K_R^a(x, t)\, \sigma_R^a(dt).$$

Given $\varepsilon > 0$, we may choose δ so that when $|\xi - t| < 2\delta$ then $|f(t) - f(\xi)| < \varepsilon$. Also

$$|(I_R^a f)(x) - f(\xi)| \leqslant \left(\int_{|\xi-t|<2\delta} + \int_{|\xi-t|\geqslant 2\delta}\right) |f(t) - f(\xi)|\, K_R^a(x, t)\, \sigma_R^a(dt)$$

The first integral, by Lemma 2.16, does not exceed ε. Also, if

$$|\xi - t| \geqslant 2\delta \text{ and } |x - \xi| < \delta, \text{ then } |x - t| \geqslant |\xi - t| - |x - \xi| > \delta$$

and so

$$K_R^a(x, t) < R^{p-2}\, \delta^{-p} (R^2 - |x-a|^2).$$

Thus, for $|x-\xi| < \delta$ the second integral does not exceed

$$2 \sup_{S_R(a)} |f(t)| (R^2-|x-a|^2) R^{p-2} \delta^{-p}$$

Since, as $x \to \xi$, $|x-a| \to R$, and since ε is arbitrary, we have (2.4) when x tends to ξ from within $B_R(a)$.

Suppose next that $x \in \mathbb{R}^p \setminus \bar{B}_R(a)$. Again by Lemma 2.16, we have

$$(I_R^a f)(x)-f(\xi) = \int\left(f(t)-\left(\frac{|x-a|}{R}\right)^{p-2} f(\xi)\right) K_R^a(x, t)\, \sigma_R^a(dt).$$

Given $\varepsilon > 0$, we may choose δ such that $|\xi-t| < 2\delta$ and $|x-\xi| < \delta$ imply that

$$\left|f(t)-\left(\frac{|x-a|}{R}\right)^{p-2} f(\xi)\right| < \varepsilon \text{ and } \left(\frac{|x-a|}{R}\right)^{p-2} < 2.$$

We split the integral as before. The first integral does not exceed ε, and the second does not exceed

$$3 \sup_{S_R(a)} |f(t)| (R^2-|x-a|^2) R^{p-2} \delta^{-p}$$

and, as before, we have the required result when x tends to ξ from without $B_R(a)$.

Theorem 2.17 shows that the Poisson integral solves the Dirichlet Problem for the ball, that is, given any continuous function f on the frontier of the ball the Poisson integral of f defines one function within the ball, another exterior to the ball, and both harmonic where they are defined which approach the given boundary values.

Furthermore, if u is harmonic in an open set $G \supset \bar{B}_R(a)$, then it coincides with $I_R^a f$ in $B_R(a)$ where $f = \operatorname{rest}_{S_R(a)} u$. For f is continuous in $S_R(a)$ and so $\lim_{x \to \xi} [(I_R^a f)(x)-u(x)] = 0$ for all $\xi \in \partial B_R(a)$ so that, by Theorem 2.10, $I_R^a f$ coincides with u in $B_R(a)$.

Expansion of a harmonic function in a power series. Suppose that u is harmonic in an open set containing $\bar{B}_R(a)$. Then, for $x \in B_R(a)$,

$$u(x) = \int u(t) K_R^a(x, t)\, \sigma_R^a(dt).$$

By (1.18), for $x \in B_R(a)$ and $t \in S_R(a)$,

$$K_R^a(x, t) = \sum_{n=0}^{\infty} \frac{2n+p-2}{p-2} R^{-n} |x-a|^n P_{n,p}(\lambda),$$

where

$$\lambda = (|x-a| R)^{-1} \sum_{i=1}^{p} (x_i-a_i)(t_i-a_i)$$

and for each such x the series is uniformly convergent in t. Hence, for $p \geqslant 3$,

$$u(x) = \sum_{n=0}^{\infty} \frac{2n+p-2}{p-2} R^{-n} \,|\, x-a \,|^n \int f(t)\, P_{n,p}(\lambda)\, \sigma_R^a(dt)$$

$$= \sum_{n=0}^{\infty} \frac{2n+p-2}{p-2} \,|\, x-a \,|^n Y_n(z), \qquad (z = (x-a)/|\, x-a \,|). \qquad (2.4)$$

By using (18′) we get a corresponding expression for $p = 2$. $Y_n(z)$ is said to be a **spherical harmonic** of order n. On the face of it, Y_n depends on R, but this is in fact not so, since, on each ray from a, z remains unchanged and the power series in $|\, x-a \,|$ must be unique.

We see then that a harmonic function is analytic, and it is because of this that we have a useful result concerning the continuation of harmonic functions.

THEOREM 2.18. *Suppose that u and v are harmonic in a domain G and that u and v coincide in a domain $G' \subset G$. Then they coincide in G.*

Let $f = u-v$. Then $f(x) = 0$ in G' and we must show that $f(x) = 0$ in G. Let

$$G'' = \{x \in G \,|\, f(y) = 0 \text{ in some neighbourhood of } x\},$$

and suppose, if possible, that G'' is a proper subset of G. Then, since G is a domain, there is $b \in \partial G'' \cap G$. Consequently, we can find $B_{2r}(b) \subset G$ and then there is an $a \in B_r(b) \cap G''$. Furthermore $b \in B_r(a) \subset G$. We may expand f about a in $B_r(a)$ in the form (2.4) and the spherical harmonics, being determined by the values of f on a sphere $S_\rho(a)$ with ρ as small as we please, will have values 0 since we may choose ρ so that $S_\rho(a) \subset G''$. Consequently, $f = 0$ in a neighbourhood of b, so that $b \in G''$—which is a contradiction. Hence $G'' = G$. This gives the result.

The Harnack Inequalities. These inequalities are rightly celebrated because, although they are trivially deduced from the Poisson integral, the consequences that may be deduced from them are both deep and powerful.

THEOREM 2.19 (*The Harnack Inequalities*). *Suppose that u is harmonic in an open set containing $\bar{B}_R(a)$ and that u is non-negative in $B_R(a)$. Then, for $x \in B_R(a)$,*

$$C_1\, u(a) \leqslant u(x) \leqslant C_2\, u(a)$$

where

$$C_1 = R^{p-2}\,(R-|\, x-a \,|)\,(R+|\, x-a \,|)^{p-1} \quad \textit{and}$$
$$C_2 = R^{p-2}\,(R+|\, x-a \,|)\,(R-|\, x-a \,|)^{p-1}$$

Since $\quad |t-a|-|x-a| \leqslant |x-t| \leqslant |t-a|+|x-a|,$

we have, setting $|t-a| = R$ and supposing $x \in B_R(a)$

$$C_1 \leqslant K_R^a(x, t) \leqslant C_2.$$

Consequently, if f is continuous and non-negative on $S_R(a)$,

$$C_1 \mathscr{S}_R^a(f) \leqslant (I_R^a f)(x) \leqslant C_2 \mathscr{S}_R^a(f).$$

If then u is as in the statement of the theorem and $f = \underset{S_R(a)}{\text{rest}}\, u$, then f is continuous and non-negative on $S_R(a)$ and $\mathscr{S}_R^a(f) = u(a)$; from which the inequalities follow.

THEOREM 2.20. *If u is harmonic and non-negative in a domain $G \subset \mathbb{R}^p$, then, given any compact $K \subset G$, there are positive constants α and β depending only on K and not on u such that, for all $x, y \in K$,*

$$\alpha < \frac{u(x)}{u(y)} < \beta.$$

First, we can find a relatively compact connected open set H such that $K \subset \bar{H} \subset G$ (the set $\{x \in G \mid \text{dist}(x, \complement G) > \rho\}$ will be connected for sufficiently small ρ) and then we need only prove the theorem for a set such as H.

So now suppose K to be the closure of a connected open set and let $\text{dist}(K_1 \subset G) = 4r$. Let $a \in K$ be fixed. Then $B_{4r}(a) \subset G$ and, setting $R = 4r$ in the Harnack Inequalities we obtain, with $C_3 = 2^{p-2}3^{1-p}$, $C_4 = 2^{p-2}3$,

$$C_3 u(a) \leqslant u(x) \geqslant C_4 u(a) \quad \text{for} \quad x \in B_{2r}(a).$$

Now $\{B_r(x)\}_{x \in K}$ is an open covering of K from which we may choose a finite covering $\{B_r(x_s)\}_{s=0,1,\ldots,n}$ with $x_0 = a$. We may suppose $x_1 \in B_{2r}(a)$ and then $B_{4r}(x_1) \subset G$ so that

$$C_3 u(x_1) \leqslant u(x) \leqslant C_4 u(x_1) \quad \text{for} \quad x \in B_{2r}(x_1).$$

In at most n steps we may pass from a to any point in K and so

$$C_3^n u(a) \leqslant u(x) \leqslant C_4^n u(a) \quad \text{for} \quad x \in K.$$

Consequently, for $x, y \in K$ and with $\alpha = C_3^{2n}$, $\beta = C_4^{2n}$ we have the required inequalities.

THEOREM 2.21. *Let $G \subset \mathbb{R}^p$ be open and let $\{u_\alpha\}_{\alpha \in A}$ be a locally uniformly bounded family of functions each of which is harmonic in G. Then $\{u_\alpha\}_{\alpha \in A}$ is locally uniformly equicontinuous in G.*

To say that the family $\{u_\alpha\}_{\alpha \in A}$ is locally uniformly bounded in G is to say that, given any compact $K \subset G$, we can find a constant λ such that, for $x \in K$ and $\alpha \in A$,

$$| u_\alpha(x) | < \lambda$$

Let $a \in G$, and suppose that $B_{3R}(a) \subset G$. Then there is a λ such that $u_\alpha(x)+\lambda > 0$ in $B_{3R}(a)$ so that $v_\alpha = u_\alpha+\lambda$ is harmonic and non-negative in $\bar{B}_{2R}(a)$. Then, for $b \in B_R(a)$, $x \in B_R(b)$, Theorem 2.20 gives

$$(C_1-1)\, v_\alpha(b) \leqslant v_\alpha(x)-v_\alpha(b) \leqslant (C_2-1)\, v_\alpha(b)$$

and, given $\varepsilon > 0$, we may choose δ, independent of α, so that, for $| x-a | < \delta$, the multiples of $v_\alpha(b)$ on the left and right are less than ε. It follows that $\{v_\alpha\}$ is uniformly equicontinuous in $B_R(a)$, and thus that the same is true for $\{u_\alpha\}$.

Hence, given $a \in G$, there is a neighbourhood of a in which the family is uniformly equicontinuous. This provides an open covering of any compact K from which we may select a finite covering, with the result that the family is uniformly equicontinuous in K, that is, given $\varepsilon > 0$ there is a $\delta > 0$ independent of α and x such that, for $\alpha \in A$ and x, $y \in K$

$$| u_\alpha(x)-u_\alpha(y) | < \varepsilon \quad \text{for } | y-x | < \delta.$$

THEOREM 2.22. *Let $G \subset \mathbb{R}^p$ be open and let $\{u_n\}$ be a locally uniformly bounded sequence of functions each of which is harmonic in G. Then,*

(*i*) *if, for $a \in G$, $\{u_n(a)\}$ converges, $\{u_n\}$ converges uniformly in some neighbourhood of a;*

(*ii*) *there is a subsequence of $\{u_n\}$ which is locally uniformly convergent in G to a harmonic function.*

Consider (*i*). Given $\varepsilon > 0$, there is $\delta > 0$, independent of n, such that

$$| u_n(x)-u_n(a) | < \tfrac{1}{3}\varepsilon \quad \text{for } | x-a | < \delta,$$

and there is N such that, for $m, n > N$,

$$| u_m(a)-u_n(a) | < \tfrac{1}{3}\varepsilon.$$

Consequently, for $| x-a | < \delta$,

$$| u_m(x)-u_n(x) | \leqslant | u_m(x)-u_m(a) | \\ +| u_m(a)-u_n(a) | +| u_n(a)-u_n(x) | < \varepsilon$$

and, by Cauchy's Convergence Criterion, (*i*) now follows.

We turn to (*ii*). Let $K \subset G$ be compact, and let $\{a_m\}$ be a sequence of points which is dense in K. From $\{u_n\}$ choose a subsequence $\{u_n^1\}$ such that $\{u_n^1(a_1)\}$ converges; from $\{u_n^1\}$ choose a subsequence such that $\{u_n^2(a_2)\}$ converges. Proceeding in this way, we find, for each m, a sequence $\{u_n^m\}$ such that $\{u_n^m(a_r)\}$ converges for $r = 1, \ldots, m$. Then the subsequence $\{v_n\}$ of $\{u_n^n\}$ given by $v_n = u_n^n$ converges for every a_m.

From (*i*), now, there is a neighbourhood of each a_m in which $\{v_n\}$ converges uniformly. This produces a (countable) open covering of K, from which we may select a finite covering and so deduce that $\{v_n\}$ converges uniformly in K.

Finally, given $a \in G$ and $B_\rho(a) \subset G$, and by Theorem 1.13,

$$\mathscr{S}_\rho^a\,(\lim_{n\to\infty} v_n) = \lim_{n\to\infty} \mathscr{S}_\rho^a(v_n) = \lim_{n\to\infty} v_n(a)$$

Also $\lim v_n$ is continuous and so it is harmonic in G.

Corollary 2.22.1. *Let $\{u_n\}$ be as Theorem 2.22. If $\{u_n\}$ converges in the neighbourhood of a point in a domain G, then it converges in G to a harmonic function.*

For if this were not the case, we could choose two different subsequences of $\{u_n\}$ each converging to a harmonic function in G. These two functions would coincide in a sub-domain of G but not everywhere in G, which, by Theorem 2.18, is impossible.

Theorem 2.23. *Let $\{u_n\}$ be a locally uniformly convergent sequence of functions each of which is harmonic in an open set G. Let D^α denote any mixed partial derivative. Then $\{D^\alpha u_n\}$ is locally uniformly convergent to $D^\alpha\,(\lim_{n\to\infty} u_n)$.*

Given $a \in G$, let $B_{2R}(a) \subset G$, and let $u(x) = \lim_{n\to\infty} u_n(x)$. Then, for $x \in B_{2R}(a)$,

$$\frac{\partial}{\partial x_i}\,[u_n(x)-u(x)] = \int [u_n(t)-u(t)]\,\frac{\partial}{\partial x_i}\,[K_{2R}^a(x, t)]\;\sigma_{2R}^a(dt)$$

and, since $\dfrac{\partial}{\partial x_i}[K_{2R}^a(x, t)] < kR^{-1-p}$ when $x \in B_R(a)$, $t \in S_{2R}(a)$ we then have

$$\left|\frac{\partial u_n}{\partial x_i}(x)-\frac{\partial u}{\partial x_i}(x)\right| < \sup_{B_{2R}(a)} |\,u_n(t)-u(t)\,|\;.\;kR^{-1-p},$$

from which it follows that $\dfrac{\partial u_n}{\partial x_i}$ converges locally uniformly to $\dfrac{\partial u}{\partial x_i}$.

Since any derivative of a harmonic function is harmonic, repetitions of this argument give the theorem.

Theorem 2.24. *Let $\{u_\alpha\}_{\alpha\in A}$ be an up-directed family of functions each of which is harmonic in a domain G. Then either $\sup_A u_\alpha$ is harmonic in G or it is identically $+\infty$ in G.*

Let

$$G_1 = \{x \in G \mid \sup_A u_\alpha(x) < +\infty\};$$

$$G_2 = \{x \in G \mid \sup_A u_\alpha(x) = +\infty\}.$$

Then we shall show that both G_1 and G_2 are open. Consequently, since $G_1 \cap G_2$ is void, $G_1 \cup G_2 = G$, and G is a domain, either G_1 or G_2 must be void.

Fix $\alpha_0 \in A$. Then, because $\{u_\alpha\}$ is up-directed, there is a $\beta_\alpha \in A$ such that $u_{\beta_\alpha} \geqslant \sup[u_{\alpha_0}, u_\alpha]$ and $\{u_{\beta_\alpha}\}_{\alpha \in A}$ is locally uniformly bounded below. Also

$$\sup_{\alpha \in A} u_\alpha \geqslant \sup_{\alpha \in A} u_{\beta_\alpha} \geqslant \sup_{\alpha \in A} u_\alpha$$

so $\sup_A u_{\beta_\alpha} = \sup_A u_\alpha$. We may therefore suppose, without loss of generality, that $\{u_\alpha\}$ is uniformly locally bounded below in G.

Suppose $a \in G_1$ and that $\bar{B}_{2R}(a) \subset G$. Then there is a constant k such that $u_\alpha(x) > k$ for $x \in \bar{B}_{2R}(a)$, and, furthermore $u_\alpha(a) \leqslant \sup_A u_\alpha(a)$. So $u_\alpha(x) - k$ is a positive harmonic function in $\bar{B}_{2R}(a)$ and so, using the right hand of Harnack's Inequalities,

$$u_\alpha(x) - k < \lambda(u_\alpha(a) - k) \text{ in } B_R(a).$$

Hence $\sup_A u_\alpha(x)$ is bounded in $B_R(a)$ and so G_1 is open.

Next, suppose $a \in G_2$. Then, using the left hand of Harnack's Inequalities, we have

$$u_\alpha(x) - k > \mu(u_\alpha(a) - k) \text{ in } B_R(a),$$

and so $\sup_A u_\alpha(x) = +\infty$ there. Thus G_2 is open.

If, now, G_2 is not void, then G_1 is; and we have the one alternative in the theorem. If, on the other hand, G_1 is not void, it coincides with G and $\sup_A u_\alpha$ is bounded above in a neighbourhood of every point of G and so bounded above in any compact $K \subset G$. Hence $\{u_\alpha\}$ is locally uniformly bounded in G and so, by Theorem 2.21, locally uniformly equicontinuous in G.

Let $u = \sup_A u_\alpha$. Given $x, y \in G$ and given $\varepsilon > 0$, there are $\alpha_1, \alpha_2 \in A$ such that $u_{\alpha_1}(x) > u(x) - \frac{1}{3}\varepsilon$ and $u_{\alpha_2}(y) > u(y) - \frac{1}{3}\varepsilon$. Since $\{u_\alpha\}$ is up-directed there is $\alpha_3 \in A$ such that $u_{\alpha_3} \geqslant \sup[u_{\alpha_1}, u_{\alpha_2}]$ and so

$$u_{\alpha_3}(x) > u(x) - \tfrac{1}{3}\varepsilon \quad \text{and} \quad u_{\alpha_3}(y) > u(y) - \tfrac{1}{3}\varepsilon.$$

Also, given $x \in G$, there is $\delta > 0$ such that

$$|u_\alpha(y) - u_\alpha(x)| < \tfrac{1}{3}\varepsilon \text{ for } |y - x| < \delta \text{ and } \alpha \in A.$$

Hence

$$| u(y)-u(x) | \leqslant | u(y)-u_{\alpha_3}(y) |+|u_{\alpha_3}(y)-u_{\alpha_3}(x) |+| u_{\alpha_3}(x)-u(x) |$$

and for $| y-x | < \delta$ this last is less than ε. Thus u is continuous in G.

Finally, using Theorem 1.18, when $B_r(a) \subset G$,

$$u(a) = \sup_A u_\alpha(a) = \sup_A \mathscr{S}_r^a(u_\alpha) = \mathscr{S}_r^a(u)$$

so that u is harmonic in G.

With this result proven, Theorem 2.25 follows immediately.

THEOREM 2.25. *Let $\{u_\alpha\}_{\alpha \in A}$ be a down-directed family of functions each of which is harmonic in a domain G. Then either* $\inf_A u_\alpha$ *is harmonic in G or it is identically $-\infty$ in G.*

§ 2.4. The Blaschke-Privalov Operators

Suppose, first, that f is twice continuously differentiable in an open set $G \subset \mathbb{R}^p$ and that $a \in G$. Then, arguing as in the proof of Pizzetti's formula, we have

$$\mathscr{S}_r^a(f)-f(a) = \frac{r^2}{2p}(\Delta f)(a)+o(r^2),$$

so that

$$(\Delta f)(a) = \lim_{r\to 0}\frac{2p}{r^2}\{\mathscr{S}_r^a(f)-f(a)\}.$$

Similarly

$$(\Delta f)(a) = \lim_{r\to 0}\frac{2(p+2)}{r^2}(\mathscr{B}_r^a(f)-f(a)).$$

These two relations prompt us to define a type of generalised Laplacian. Suppose that f is σ_r^a-integrable for all small r. Then we define the spherical Blaschke-Privalov operators $\bar{S}$ and $\underline{S}$ by

$$(\bar{S}f)(a) = \limsup_{r\to 0}\frac{2p}{r^2}\{\mathscr{S}_r^a(f)-f(a)\}$$

$$(\underline{S}f)(a) = \liminf_{r\to 0}\frac{2p}{r^2}\{\mathscr{S}_r^a(f)-f(a)\}$$

and the ball Blaschke-Privalov operators $\bar{B}$ and $\underline{B}$ by

$$(\bar{B}f)(a) = \limsup_{r\to 0}\frac{2(p+2)}{r^2}\{\mathscr{B}_r^a(f)-f(a)\}$$

$$(\underline{B}f)(a) = \liminf_{r\to 0}\frac{2(p+2)}{r^2}\{\mathscr{B}_r^a(f)-f(a)\}$$

when f is locally integrable in G.

Then, if f is l.s.c. in G, we have

THEOREM 2.26. $(\underline{S}f)(a) \leqslant (\underline{B}f)(a) \leqslant (\bar{B}f)(a) \leqslant (\bar{S}f)(a)$.

We have

$$r^p \mathscr{B}_r^a(f) = p \int \rho^{p-1} \mathscr{S}_r^a(f)\, d\rho$$

and so

$$r^p (\mathscr{B}_r^a(f) - f(a)) = \tfrac{1}{2} \int_0^r \rho^{p+1} \left\{ \frac{2p}{\rho^2} (\mathscr{S}_\rho^a(f) - f(a)) \right\} d\rho$$

which is

$$\leqslant \frac{1}{2} \frac{r^{p+2}}{p+2} \sup_{0 \leqslant \rho \leqslant r} \frac{2p}{\rho^2} (\mathscr{S}_\rho^a(f) - f(a))$$

Hence

$$(\bar{B}f)(a) = \limsup_{r \to 0} \frac{2(p+2)}{r^2} (\mathscr{B}_r^a(f) - f(a))$$

$$\leqslant \limsup_{\rho \to 0} \frac{2p}{\rho^2} (\mathscr{S}_\rho^a(f) - f(a)) = (\bar{S}f)(a).$$

A similar argument, using infimum in place of supremum and inequalities in the opposite direction, gives the left-hand inequality.

These operators are important, because they enable us to state necessary and sufficient conditions for the super- and subharmonicity of semicontinuous functions. Before we do this, we first prove a necessary preliminary which is, however, an important result in its own right.

THEOREM 2.27. *Let f be l.s.c. (u.s.c.) in $G \subset \mathbb{R}^p$. Then f is super- (sub-) harmonic in G if, and only if, whenever $\bar{B}_r(a) \subset G$, and for $x \in B_r(a)$,*

$$f(x) \geqslant (I_r^a f)(x) \quad (f(x) \leqslant (I_r^a f)(x)). \tag{2.5}$$

Suppose f superharmonic. Let $g \in \mathscr{C}(S_r(a))$ and be such that $g(\xi) \leqslant f(\xi)$ in $S_r(a)$. Then

$$\liminf_{x \in B_r(a),\, x \to \xi} [f(x) - (I_r^a g)(x)] \geqslant f(\xi) - g(\xi),\ \xi \in S_r(a)$$

since f is l.s.c. and by Theorem 2.17. Hence, by Theorem 2.4

$$f(x) \geqslant (I_r^a g)(x),\ x \in B_r(a).$$

But, by Theorem 1.17, $(I_r^a f)(x) = \sup_{g \leqslant f} (I_r^a g)(x)$, and so (2.5) follows for superharmonic functions. A similar argument will give (2.5) for subharmonic functions.

Conversely, suppose (2.5) holds. Then, setting $x = a$, we have $f(a) \geqslant \mathscr{S}_r^a(f)$ $(f(a) \leqslant \mathscr{S}_r^a(f))$, and so f is super- (sub-) harmonic.

THEOREM 2.28. *Suppose that f is l.s.c. in $G \subset \mathbb{R}^p$. Then a necessary and sufficient condition that f be superharmonic in G is that either*

$$(\underline{S}f)(a) \leqslant 0 \text{ for all } a \in G \tag{2.6}$$

or

$$(\underline{B}f)(a) \leqslant 0 \text{ for all } a \in G \tag{2.7}$$

If f is superharmonic in G then $\mathscr{S}_r^a(f) \leqslant f(a)$ (and $\mathscr{B}_r^a(f) \leqslant f(a)$) for suitable and, indeed, all sufficiently small r, and so clearly (2.6) and (2.7) hold.

Suppose, conversely, that (2.6) is satisfied. It is then enough to show, by Theorem 2.27, that for $\bar{B}_r(a) \subset G$ we have

$$(I_r^a f)(x) \leqslant f(x),\; x \in B_r(a);$$

and to show this, it is enough to show, by Theorem 1.17, that, for every $g \in \mathscr{C}(S_r(a))$ such that $g \leqslant f$,

$$(I_r^a g)(x) \leqslant f(x),\; x \in B_r(a) \tag{2.8}$$

The function $w(x)$ given by $w(x) = |x-a|^2 - r^2$ has $(\Delta w)(x) = 2p$. Furthermore, $w(x) = 0$ when $x \in S_r(a)$. Let

$$v(x) = (I_r^a g)(x) - f(x) + \varepsilon\, w(x).$$

Then (2.8) will hold if

$$v(x) \leqslant 0 \text{ for every } \varepsilon > 0. \tag{2.9}$$

Now v is u.s.c. in $B_r(a)$. Suppose, if possible, that (2.9) does not hold Then v would attain a positive maximum in $B_r(a)$ at some point b, say. Then

$$(\bar{S}v)(b) = \varepsilon\,(\Delta w)(b) - (\underline{S}f)(b) = 2p\varepsilon - (\underline{S}f)(b) > 0.$$

But then there would be an $r > 0$ such that $\mathscr{S}_r^b(v) > v(b)$, and this would be incompatible with v having a local maximum at b. Thus (2.9) holds, whence (2.8) and (2.6).

Now suppose that (2.7) holds. Then, by Theorem 2.26, (2.6) holds and so f is superharmonic in G.

It will now follow immediately that

THEOREM 2.29. *Let f be u.s.c. in $G \subset \mathbb{R}^p$. Then a necessary and sufficient condition that f be subharmonic in G is that either*

$$(\bar{S}f)(a) \geqslant 0 \text{ for } a \in G$$

or

$$(\bar{B}f)(a) \geqslant 0 \text{ for } a \in G.$$

From Theorems 2.28 and 2.29 we may deduce the following criterion for super- (sub-) harmonicity:

A function f defined in $G \subset \mathbb{R}^p$ is super- (sub-) harmonic in G if, and only if,

(*a*) *f is l.s.c. (u.s.c.) in G.*

(*b*) *Given $a \in G$, there is $r_0 > 0$ such that, for $r \leqslant r_0$,*

$$\mathscr{S}_r^a(f) \leqslant f(a) \; (\mathscr{S}_r^a(f) \geqslant f(a))$$

Furthermore, *in the above $\mathscr{S}_r^a$ may be replaced by $\mathscr{B}_r^a$*. Thus, we now see that the implication (1) deduced in the proof of Theorem 2.2, that is, if f is superharmonic it is 'super ball mean' is, in fact, an equivalence.

§ 2.5. Approximation by Smooth Functions

In this section we discuss the approximation of superharmonic functions by superharmonic functions as smooth as we please.

THEOREM 2.30. *Let f be superharmonic in $G \subset \mathbb{R}^p$ and suppose that $\bar{B}_r(a) \subset G$. Then $\mathscr{S}_r^a(f)$ and $\mathscr{B}_r^a(f)$ are decreasing functions of r, are supermean functions of a in the open set in which they are defined, and*

$$\lim_{r\to 0} \mathscr{S}_r^a(f) = \lim_{r\to\infty} \mathscr{B}_r^a(f) = f(a).$$

Let $0 < \rho < r$. Now, by Theorem 2.27,

$$\mathscr{S}_\rho^a(f) = \int f(t)\, \sigma_\rho^a(dt) = (I_\rho^a f)\,(a) \leqslant f(a).$$

Also

$$f(t) \geqslant (I_r^a f)\,(t) \text{ for } t \in S_\rho(a), \text{ and so}$$

$$(I_\rho^a f)\,(x) \geqslant I_\rho^a[(I_r^a f)]\,(x) \text{ for } x \in B_\rho(a)$$

and, in particular,

$$\mathscr{S}_\rho^a(f) = (I_\rho^a f)\,(a) \geqslant \mathscr{S}_\rho^a[I_r^a f] = (I_r^a f)\,(a)$$

since $I_r^a f$ is harmonic in $\bar{B}_\rho(a)$, and $(I_r^a f)\,(a) = \mathscr{S}_r^a(f)$ showing that $\mathscr{S}_r^a(f)$ is a decreasing function of r.

Next

$$\mathscr{B}_r^a(f) = \frac{p}{r^p} \int_0^r \rho^{p-1}\, \mathscr{S}_\rho^a(f)\, d\rho = p \int_0^1 u^{p-1}\, \mathscr{S}_{ru}^a(f)\, du,$$

and since $\mathscr{S}_{ru}^a(f)$ is a decreasing function of r, it follows that $\mathscr{B}_r^a(f)$ is also decreasing. Also

$$\mathscr{B}_r^a(f) \geqslant p \int_0^1 u^{p-1}\, \mathscr{S}_r^a(f)\, du = \mathscr{S}_r^a(f).$$

Next, letting $g(x) = \mathscr{S}_\rho^x(f)$, we have

$$\mathscr{S}_r^a(g) = \int g(a+x)\, \sigma_r^0(dx) = \iint f(a+x+t)\, \sigma_\rho^0(dt)\, \sigma_r^0(dx).$$

Inverting the order of integration, by Fubini's Theorem, this last may be rewritten as $\int \mathscr{S}_r^{a+t}(f)\, \sigma_\rho^0(dt)$ and this, since f is superharmonic, is not greater than $\int f(a+t)\, \sigma_\rho^0(dt)$.

Thus $\mathscr{S}_r^a(g) \leqslant g(a)$ showing that $\mathscr{S}_\rho^x(f)$ is supermean in the open subset of G where it is defined. A similar argument shows the same result for $\mathscr{B}_\rho^x(f)$.

Finally, since f is l.s.c. in G, given $\varepsilon > 0$, there is $\delta > 0$ such that, when $|x-a| < \delta, f(x) > f(a)-\varepsilon$ and hence, for $\rho < \delta$,

$$\mathscr{S}_\rho^a(f) > \int (f(a)-\varepsilon)\, \sigma_\rho^a(dt) = f(a)-\varepsilon.$$

Also $\mathscr{S}_\rho^a(f) \leqslant f(a)$, and hence $\lim\limits_{\rho\to 0} \mathscr{S}_\rho^a(f) = f(a)$. Also

$$\mathscr{S}_r^a(f) \leqslant \mathscr{B}_r^a(f) \leqslant f(a);$$

and so $\lim\limits_{r\to 0} \mathscr{B}_r^a(f) = f(a)$.

It now follows immediately that

THEOREM 2.31. *Let f be subharmonic in G and suppose that $\bar{B}_r(a) \subset G$. Then $\mathscr{S}_r^a(f)$ and $\mathscr{B}_r^a(f)$ are increasing functions of r, submean functions of a where they are defined, and*

$$\lim_{r\to 0} \mathscr{S}_r^a(f) = \lim_{r\to 0} \mathscr{B}_r^a(f) = f(a).$$

COROLLARY 2.32. *Suppose that f and g are superharmonic in G and that $f(x) = g(x)$ a.e. in G. Then $f(x) = g(x)$ in G.*

Given $a \in G$ and sufficiently small r, $\mathscr{B}_r^a(f) = \mathscr{B}_r^a(g)$; whence, by Theorem 2.30, $f(a) = g(a)$.

THEOREM 2.33. *Suppose that f is twice continuously differentiable in $G \subset \mathbb{R}^p$. Then f is supermean in G if, and only if,*

$$(\Delta f)(x) \leqslant 0 \text{ for } x \in G.$$

As in the proof of Theorem 2.13, when $\bar{B}_R(a) \subset G$,

$$\int_{BR(a)} \Delta f\, dx = R^{p-1} \frac{2\pi^{\frac{1}{2}p}}{\Gamma(\frac{1}{2}p)} \frac{d}{dR} (\mathscr{S}_R^a(f)).$$

Suppose that f is superharmonic in G. Then, by Theorem 2.30, the right hand side is non-positive, and thus the left hand side is so for all $\bar{B}_R(a) \subset G$. Consequently, $(\Delta f)(x) \leqslant 0$.

Conversely, suppose that $\Delta f \leqslant 0$ in G. Then $\mathscr{S}_R^a(f)$ is decreasing in R and, since f is continuous, $\lim\limits_{r\to 0} \mathscr{S}_R^a(f) = f(a)$. Hence $f(a) \geqslant \mathscr{S}_R^a(f)$ and so f is superharmonic in G.

Similarly,

THEOREM 2.34. *Suppose that f is twice continuously differentiable in $G \subset \mathbb{R}^p$. Then f is subharmonic in G if, and only if,*

$$(\Delta f)(x) \geqslant 0 \text{ in } G.$$

Given an open set $G \subset \mathbb{R}^p$, define the open set G_r by

$$G_r = \{x \in G \mid \text{dist}(x, G) > r\}.$$

THEOREM 2.35. *If f is locally integrable in G then $\mathscr{B}_r^x(f)$ is continuous in G_r. If f is n times continuously differentiable in G then $\mathscr{B}_r^x(f)$ is $n+1$ times continuously differentiable in G_r.*

We have

$$\frac{\pi^{\frac{1}{2}p} r^p}{\Gamma(\frac{1}{2}p+1)} \left| \mathscr{B}_r^{x+h}(f) - \mathscr{B}_r^x(f) \right| \leqslant \int_{B_r(x+h) \,\Delta\, B_r(x)} |f(t)| \, dt$$

where Δ is the symmetric difference operation defined by

$$M \,\Delta\, N = (M \cup N) \setminus (M \cap N).$$

The measure of $B_r(x+h) \,\Delta\, B_r(x)$ is given by

$$\frac{2\pi^{\frac{1}{2}p} r^p}{\Gamma(\frac{1}{2}p+1)} - \frac{\pi^{\frac{1}{2}(p-1)}}{\Gamma(\frac{1}{2}p+\frac{1}{2})} \left\{ \int_{|h|-r}^{\frac{1}{2}|h|} (r^2 - (t-|h|)^2)^{\frac{p-1}{2}} \, dt + \int_{\frac{1}{2}|h|}^{r} (r^2 - t^2)^{\frac{p-1}{2}} \, dt \right\}$$

and, as $h \to 0$, this tends to zero. Hence the integral on the right tends to zero as $h \to 0$ and so $\mathscr{B}_r^x(f)$ is continuous.

Assume now that f is n times continuously differentiable in G. Then, by Theorem 1.25, for any derivative of order not greater than n, $D^\alpha(\mathscr{B}_r^x(f)) = \mathscr{B}_r^x(D^\alpha f)$ so that $\mathscr{B}_r^x(f)$ is n times differentiable in G_r. To prove that $\mathscr{B}_r^x(f)$ is $(n+1)$ times differentiable, it is enough to show that, if f is continuous in G, then $\dfrac{\partial}{\partial x_i}(\mathscr{B}_r^x(f))$ exists and is continuous in G^T, for $i = 1, \ldots, p$. Without loss of generality we may suppose $i = 1$.

Make the transformation (regarding r as fixed)

$$\begin{aligned} t_1 &= \lambda + r\cos\theta_1 \\ t_i &= r\sin\theta_1 \ldots \sin\theta_{i-1}\cos\theta_i \quad (i = 2, \ldots, p-1) \\ t_p &= r\sin\theta_1 \ldots \sin\theta_{p-1}. \end{aligned}$$

Then

$$\frac{\partial(t_1, \ldots, t_p)}{\partial(\lambda, \theta_1, \ldots, \theta_{p-1})} = r^{p-1}\cos\theta_1 \sin^{p-2}\theta_1 \ldots \sin\theta_{p-2} = J, \text{ say,}$$

and, with $h = (h_1, 0, \ldots, 0)$,

$$\frac{\pi^{\frac{1}{2}p}\, r^p}{\Gamma(\frac{1}{2}p+1)}\, \frac{1}{h_1}\, \{\mathscr{B}_r^{x+h}(f) - \mathscr{B}_r^x(f)\}$$

$$= \frac{1}{h_1} \int_0^{h_1} \int_0^{\arccos \frac{h_1+\lambda}{2r}} \int_0^{\pi} \cdots \int_0^{\pi} \int_0^{2\pi} f(x+t)\, J\, d\theta_{p-1} \cdots d\theta_1\, d\lambda$$

$$- \frac{1}{h_1} \int_0^{h_1} \int_{\arccos \frac{\lambda}{2r}}^{\pi} \int_0^{\pi} \cdots \int_0^{\pi} \int_0^{2\pi} f(x+t)\, J\, d\theta_{p-1} \cdots d\theta_1\, d\lambda$$

and, since f is continuous, the right hand side converges as $h_1 \to 0$, and we have

$$\frac{\partial}{\partial x_1}(\mathscr{B}_r^x(f)) = \left\{\int_0^{\frac{1}{2}\pi} \int_0^{\pi} \cdots \int_0^{\pi} \int_0^{2\pi} - \int_{\frac{1}{2}\pi}^{\pi} \int_0^{\pi} \cdots \int_0^{\pi} \int_0^{2\pi}\right\}$$

$$f(x+t)\, J\, d\theta_{p-1} \cdots d\theta_1.$$

Furthermore, since f is continuous, the integral on the right is a continuous function of x.

This done, we may now produce an approximation theorem.

THEOREM 2.36. *Suppose that f is superharmonic (subharmonic) in G. Then, given $a \in G$, there is an increasing (decreasing) sequence $\{f_k\}$ of functions with f_k superharmonic (subharmonic) in an open set $H \subset G$ containing a, with f_k n times continuously differentiable in H and such that* $\lim_{k\to\infty} f_k(a) = f(a)$.

Suppose that f is superharmonic in G. Let $a \in G$, and suppose that $B_R(a) \subset G$. Then, if $nk^{-1} < \frac{1}{2}R$ and $x \in G_{\frac{1}{2}R}$, g_k^i is inductively defined for $i = 1, \ldots, n$ by

$$g_k^0(x) = \mathscr{B}_{1/k}^x(f); \qquad g_k^i(x) = \mathscr{B}_{1/k}^x(g_k^{i-1}),$$

and $f_k = g_k^n$ is n times continuously differentiable in $G_{\frac{1}{2}R}$.

Then $\{f_k\}$ is an increasing sequence of superharmonic functions, by Theorem 2.30. Furthermore, $g_h^0(a)$ converges increasingly to $f(a)$. Suppose that $g_k^r(a)$ converges increasingly to $f(a)$. Now

$$g_k^{r+1}(a) \geqslant \mathscr{B}_{1/k}^a(g_l^r)$$

when $k \geqslant l$, since g_l^r is increasing in l. Hence, using Theorem 2.30 again,

$$f(a) \geqslant \lim_{k\to\infty} g_k^{r+1}(a) \geqslant g_l^r(a)$$

for every l and consequently $g_k^{r+1}(a)$ converges increasingly to $f(a)$. Thus $f_k(a)$ also does so.

The result for subharmonic functions now follows in the usual way.

Thus, any superharmonic function can be approximated by an increasing sequence of superharmonic functions which are finitely as many times continuously differentiable as we please. But it is not possible by this method to find infinitely differentiable approximating superharmonic functions. This goal can be attained in a different way, however.

THEOREM 2.37. *Suppose that f is as in Theorem 2.36. Then, given $a \in G$, there is a sequence $\{f_k\}$ of functions with f_k superharmonic (subharmonic) in an open set $H \subset G$ containing a with f_k infinitely differentiable and such that* $\lim_{k\to\infty} f_k(a) = f(a)$.

Suppose that f is superharmonic in G and, in the first instance, that f is continuous. Let

$$g_k = f * \tau_{1/k} \text{ (τ_δ is defined in Chapter 1, §4).}$$

Then

$$\begin{aligned}\mathscr{B}_r^a(g_k) &= \int\int f(t)\, \tau_{1/k}(x-t)\, dt\, \beta_r^a(dx) \\ &= \int \tau_{1/k}(t) \int f(x+t)\, \beta_r^a(dx)\, dt \leqslant \int f(a+t)\, \tau_{1/k}(t)\, dt = g_k(a),\end{aligned}$$

and g_k is infinitely differentiable in $G_{1/k}$, and so continuous. It is therefore superharmonic there and, by Theorem 1.29 (*ii*), $\lim_{k\to\infty} g_k(a) = f(a)$.

Now, more generally, suppose that f is superharmonic. Then $f_m = \mathscr{B}_{1/m}^a(f)$ is continuous and superharmonic and $\lim_{m\to\infty} f_m(a) = f(a)$. Setting $g_{mk} = f_m * \tau_{1/k}$, we now can find a sequence $\{k_m\}$ such that g_{mk_m} converges to f as m tends to ∞, and g_{mk_m} is superharmonic and infinitely differentiable.

There is a similar result for subharmonic functions.

Theorem 2.36, applied to a harmonic function, shows that each f_k constructed there is equal to f, and so f is n times continuously differentiable for every n. This provides another, independent, proof that a harmonic function is infinitely differentiable.

§ 2.6. The Riesz Decomposition

We have seen that, if f is a function smooth enough to possess a continuous Laplacian Δf, then $\Delta f \leqslant 0$ is a necessary and sufficient condition that f be superharmonic. We shall now define a generalised Laplacian which will have an existence even when f is not smooth. But, for this extra generality a price must be exacted, and it is this: **Δf can no longer be a point function;** on the contrary, it must be a functional defined on $\mathscr{D}(G)$.

Given a function f which is locally integrable on an open set $G \subset \mathbb{R}^p$, we define Δf by

$$(\Delta f)(\phi) = \int f(x)(\Delta\phi)(x)\,dx \text{ for } \phi \in \mathcal{D}(G).$$

We note that if f is continuously twice differentiable then

$$(\Delta f)(\phi) = \int f(x)(\Delta\phi)(x)\,dx$$

$$= \sum_{i=1}^{p} \int_{-\infty}^{\infty} \cdots \int_{-\infty}^{\infty} f(x) \frac{\partial^2 \phi}{\partial x_i^2}\,dx_i\,dx_1 \ldots dx_{i-1}\,dx_{i+1} \ldots dx_p,$$

and the inmost integral becomes, on integrating by parts twice and recalling that ϕ has compact support,

$$\int \phi(x) \frac{\partial^2 f}{\partial x_i^2}\,dx_i,$$

so that

$$(\Delta f)(\phi) = \int (\Delta f)(x)\,\phi(x)\,dx, \tag{2.10}$$

where Δ is used in the generalised sense on the left and in the classical sense on the right.

THEOREM 2.38. *Suppose that g is locally integrable in an open set $G \subset \mathbb{R}^p$. Then a necessary and sufficient condition that g be equal to a superharmonic function a.e. in G is that*

$$(\Delta g)(\phi) \leqslant 0 \textit{ for all } \phi \in \mathcal{D}(G) \textit{ with } \phi \geqslant 0.$$

Suppose first that g is equal a.e. to a function f superharmonic in G. Then $(\Delta f)(\phi) = \Delta g(\phi)$ for all $\phi \in \mathcal{D}(G)$. By Theorem 2.36 there is defined in supp ϕ a sequence $\{f_n\}$ of functions each of which is superharmonic, such that f_n is continuously twice differentiable and f_n converges increasingly to f. Then, by Beppo Levi,

$$(\Delta f)(\phi) = \int f(t)(\Delta\phi)(t)\,dt = \lim_{n\to\infty} \int f_n(t)\,\Delta\phi(t)\,dt =$$

$$\lim_{n\to\infty} \int (\Delta f_n)(t)\,\phi(t)\,dt.$$

But, by Theorem 2.33, $(\Delta f_n)(t) \leqslant 0$, and so the last integral is non-positive for every non-negative ϕ. Consequently, $\Delta f \leqslant 0$ and hence $\Delta g \leqslant 0$.

Conversely, suppose that $\Delta g \leqslant 0$. Let $g_\delta = g * \tau_\delta$. Then

$$(\Delta g_\delta)(x) = \Delta \left(\int g(t)\,\tau_\delta(x-t)\,dt = \int g(t)\,\Delta\tau_\delta(x-t)\,dt\right.$$

and this last is non-positive since $\tau_\delta \geqslant 0$ and Δg is a non-positive functional. Hence, by Theorem 2.33, g_δ is superharmonic and, by Theorem 1.29 (*iv*),

$$\mathcal{B}_r^x(g) = \lim_{\delta\to 0} \mathcal{B}_r^x(g_\delta)$$

By Theorem 2.30, $\mathscr{B}_r^x(g_\delta)$ is a decreasing function of r and so, therefore, is $\mathscr{B}_r^x(g)$. Consequently,

$$f(x) = \lim_{r \to 0} \mathscr{B}_r^x(g)$$

exists for all $x \in G$, and $f(x) \geqslant \mathscr{B}_r^x(g)$ when $r > 0$.

Now, by Theorem 1.26, $\lim_{r \to 0} \mathscr{B}_r^x(g) = g(x)$ a.e. in G, so that $f(x) = g(x)$ a.e. in G and hence $\mathscr{B}_r^x(f) = \mathscr{B}_r^x(g)$. Thus

$$f(x) \geqslant \mathscr{B}_r^x(f) \text{ for } x \in G.$$

Furthermore, since $\mathscr{B}_r^x(g)$ is continuous, f is l.s.c. by Theorem 1.6, and thus f is superharmonic in G. Hence g is equal to a superharmonic function a.e. in G.

THEOREM 2.39. *Suppose that g is locally integrable in open $G \subset \mathbb{R}^p$. Then a necessary and sufficient condition that g be equal to a subharmonic function a.e. in G is that*

$$(\Delta g)(\phi) \geqslant 0 \textit{ for all } \phi \in \mathscr{D}(G) \textit{ such that } \phi \geqslant 0.$$

This follows immediately from Theorem 2.38.

THEOREM 2.40. *Suppose that g is locally integrable in open $G \subset \mathbb{R}^p$. Then a necessary and sufficient condition that g be equal to a harmonic function a.e. in G is that*

$$\Delta g(\phi) = 0 \textit{ for all } \phi \in \mathscr{D}(G).$$

If g is equal a.e. to a harmonic function f in G, then by (2.10) $(\Delta f)(\phi) = 0$ and $(\Delta g)(\phi) = (\Delta f)(\phi)$, and so the necessity follows.

If $(\Delta g)(\phi) = 0$ for all $\phi \in \mathscr{D}(G)$ then the sufficient condition in Theorems 2.38 and 2.39 hold, and so g is equal a.e. to both a superharmonic function f_1 and a subharmonic function f_2.

Consequently $f_1 = f_2$ a.e. Now $f_1 - f_2$ is superharmonic in G and so l.s.c. there. Hence, if $f_1(a) > f_2(a)$, this inequality will hold in a neighbourhood of a—a set of positive measure—which is impossible. Hence $f_1 - f_2 \leqslant 0$ in G. But for $a \in G, f_1(a) \geqslant \mathscr{B}_r^a(f_1) = \mathscr{B}_r^a(f_2) \geqslant f_2(a)$. Thus $f = f_2 = f_2$ is harmonic in G.

Suppose that $Y = \mathbb{R}^p$ when $p \geqslant 3$, and $Y = B_1(0)$ when $p = 2$. Given any Radon measure μ on Y, we define the **Newtonian potential** V_μ of μ by

$$V_\mu(x) = \int h(x-t)\, \mu(dt).$$

We shall suppose also that, when $p = 2$, supp μ is compact. By Fatou's Lemma (Theorem 1.24)

$$\liminf_{y \to x} V_\mu(y) \geqslant \int \liminf_{y \to x} h(y-t)\, \mu(dt) = \int h(x-t)\, \mu(dt) = V_\mu(x)$$

so that V_μ is l.s.c. in Y.

As we shall see later (Theorem 3.4) a necessary and sufficient condition that V_μ be finite a.e. in $\mathbb{R}^p$ ($p \geq 3$) is that

$$\int_{|y|\geq 1} |y|^{2-p} \mu(dy) < +\infty$$

If this does not hold, V_μ is identically $+\infty$ in $\mathbb{R}^p$, and we shall suppose that the condition is satisfied. It certainly is, if supp μ is compact.

We have

$$\mathscr{S}_r^a(V_\mu) = \iint h(x-t)\, \mu(dt)\, \sigma_r^a(dx),$$

and we may, by Fubini's Theorem, invert the order of integration. By Lemma 2.15 we then obtain

$$\mathscr{S}_r^a(V_\mu) = \int_{B_r(a)} h(r)\, \mu(dt) + \int_{\complement B_r(a)} h(t-a)\, \mu(dt)$$

and thus

$$V_\mu(a) - \mathscr{S}_r^a(V_\mu) = \int_{B_r(a)} (h(t-a)-h(r))\, \mu(dt)$$

Now, if $a \notin$ supp μ, we can find $r > 0$ such that $\mu(B_r(a)) = 0$, and then

$$V_\mu(a) = \int_{\complement B_r(a)} h(t-a)\, \mu(dt) = \mathscr{S}_r^a(V_\mu)$$

Also, noting that $h(x-t) \leq h(\frac{1}{2}r)$ when $x \in B_{\frac{1}{2}r}(a)$, $t \in \complement B_r(a)$, the continuity of V_μ follows from Theorem 1.23. Thus V_μ is harmonic at a.

If $a \in$ supp μ then $\mu(B_r(a)) \neq 0$ for every $r > 0$ and so, by Lemma 1.12, $V_\mu(a) > \mathscr{S}_r^a(V_\mu)$.

Thus, summing up, **V_μ is superharmonic in Y, is harmonic in $Y \setminus$ supp μ, and is not harmonic in supp μ.**

We turn now to the result which gives this section its title. It is a sort of converse to what we have just said about the potential of a measure with a compact support.

THEOREM 2.41 (*The local form of the Riesz Decomposition*). *Suppose that f is superharmonic in an open set $G \subset Y$. Let the measure μ be defined on G by*

$$\mu(\phi) = C_p\, (\Delta f)\, (\phi),\ \phi \in \mathscr{D}(G)$$

where $C_p = -\Gamma(\frac{1}{2}p)/(p-2)\, 2\pi^{\frac{1}{2}p}$ when $p \geq 3$ and $C_2 = -1/2\pi$. Then, given any open $H \subset \bar{H} \subset G$, there is a function g harmonic in H such that, for $x \in H$,

$$f(x) = \int_H h(x-t)\, \mu(dt) + g(x).$$

Let μ_H be the restriction of μ to H, that is, $\mu_H(E) = \mu(E \cap H)$. Then V_{μ_H} is locally integrable and

$$(\Delta V_{\mu_H})\, (\phi) = \int V_{\mu_H}(x)\, (\Delta\phi)\, (x)\, dx = \iint h(x-t)\, \mu_H(dt)\, (\Delta\phi)\, (x)\, dx$$

and, by Fubini's Theorem, we may invert the order of integration. The inner integral, then, has the value $C_p^{-1}\,\phi(t)$, by (2.3), and so

$$(\Delta V_{\mu_H})\,(\phi) = C_p^{-1}\,\mu_H(\phi) = (\Delta f)\,(\phi) \text{ for supp } \phi \subset H.$$

Thus $\Delta(V_{\mu_H}-f)\,(\phi) = 0$ for all $\phi \in \mathscr{D}(G)$ with supp $\phi \subset H$ and so, by Theorem 2.40, $V_{\mu_H}-f$ is equal a.e. to a harmonic function. Hence setting $v_n(x) = B^x_{1/n}(V_{\mu_H})$, $f_n(x) = B^x_{1/n}(f)$ we have $v_n(x)-f_n(x) = g(x)$. Since v_n and f_n converge to V_{μ_H} and f respectively we now have $V_{\mu_H}-f = g$ except where both terms on the left are $+\infty$, in which case the left hand side is undefined. It is now natural to define the value of the lefthand side to be $g(x)$ whenever this happens. With this gloss the theorem now follows.

§ 2.7. Expansion of a Function Harmonic in an Annulus

Suppose that $\bar{A} = \bar{B}_R(a) - B_r(a)$, with $R > r > 0$ and that f is harmonic in an open set containing $\bar{A}$. Let $y \in A$, and set

$$g_y(x) = \int h(x-y-t)\,\tau_\varepsilon\,(t)\,dt.$$

Then g_y is infinitely differentiable and so, by Lemma 2.12,

$$\int_A f(x)\,\Delta g_y(x)\,dx =$$

$$\frac{2\pi^{\frac{1}{2}p}}{\Gamma(\frac{1}{2}p)}\left\{R^{p-1}\int\left(f(x)\frac{\partial g_y}{\partial R}-g_y\frac{\partial f}{\partial R}\right)\sigma^a_R(dx)\right.$$

$$\left.-r^{p-1}\int\left(f(x)\frac{\partial g_y}{\partial r}-g_y\frac{\partial f}{\partial r}\right)\sigma^a_r(dx)\right\}, \tag{2.11}$$

where $\frac{\partial}{\partial R}, \frac{\partial}{\partial r}$ denote partial differentiation with respect to $|\,x-a\,|$ taken at values $|\,x-a\,| = R$ and $|\,x-a\,| = r$ respectively. Now

$$\Delta_x g_y(x) = \int h(t)\,\Delta_x\,\tau_\varepsilon\,(x-y-t) = \int h(t)\,\Delta_t\,\tau_\varepsilon\,(x-y-t)\,dt$$
$$= \int h(t)\,\Delta_t\,\tau_\varepsilon\,(t+y-x)\,dt = \int h(t+x-y)\,\Delta\tau_\varepsilon\,(t)\,dt$$

and this, by (2.3), is equal to $C_p\,\tau_\varepsilon\,(x-y)$, where

$$C_p = (2-p)\,2\pi^{\frac{1}{2}p}/\Gamma(\tfrac{1}{2}p) \text{ when } p \geqslant 3 \text{ and } C_2 = -2\pi.$$

Hence $C_p \int f(x)\,\tau_\varepsilon\,(x-y)\,dx$ is equal to the right-hand side of (2.11).

Since $y \notin \text{supp } \sigma^a_R \cup \text{supp } \sigma^a_r$ we have that

$$\frac{\partial g_y}{\partial R} \to \frac{\partial h}{\partial R}, \frac{\partial g_y}{\partial r} \to \frac{\partial h}{\partial r} \text{ and } g_y \to h$$

uniformly in x as $\varepsilon \to 0$ and so, letting $\varepsilon \to 0$, we obtain, denoting $h(x-y)$ by $h_y(x)$,

$$f(y) = \frac{1}{2-p}\left\{R^{p-1}\int\left(f(x)\frac{\partial h_y}{\partial R} - h_y(x)\frac{\partial f}{\partial R}\right)\sigma_R^a(dx)\right.$$
$$\left. - r^{p-1}\int\left(f(x)\frac{\partial h_y}{\partial r} - h_y(x)\frac{\partial f}{\partial r}\right)\sigma_r^a(dx)\right\} \tag{2.12}$$

when $p \geqslant 3$ and, when $p = 2$, $2-p$ is replaced by -1 in the right-hand side.

Now, by (1.17) with $\rho = |y-a| < |x-a| = R$ we have

$$h_y(x) = h_a(x) + \sum_{n=1}^{\infty} \rho^n R^{2-p-n} P_{n,p}(\lambda_1)$$

where

$$R\rho\lambda_1 = \sum_{i=1}^{p} (x_i - a_i)(y_i - a_i)$$

and hence

$$\frac{\partial}{\partial R}(h_y) = \sum_{n=0}^{\infty} (2-p-n)\,\rho^n R^{1-p-n} P_{n,p}(\lambda_1)$$

(it is to be noted that λ_1 is, in fact, independent of R) when $p \geqslant 3$ and a like result when $p = 2$ except that the first term is replaced by $-1/R$.

When, on the other hand, $\rho = |y-a| > |x-a| = r$, we have

$$h_y(x) = h_a(y) + \sum_{n=1}^{\infty} r^n \rho^{2-p-n} P_{n,p}(\lambda_2)$$

where

$$r\rho\lambda_2 = \sum_{i=1}^{p} (x_i - a_i)(y_i - a_i)$$

and hence

$$\frac{\partial}{\partial r}(h_y) = \sum_{n=1}^{\infty} n r^{n-1} \rho^{2-p-n} P_{n,p}(\lambda_2).$$

Consequently

$$f(y) = K + \alpha h_a(y) + \sum_{n=1}^{\infty} Y_n \rho^n + \sum_{n=1}^{\infty} Z_n \rho^{2-p-n}$$

where

$$(2-p)\,Y_n = R^{-n}\int P_{n,p}(\lambda_1)\left[(2-p-n)f(x) - R\frac{\partial f}{\partial R}\right]\sigma_R^a(dx) \tag{2.13}$$

$$(2-p)\,Z_n = r^{n+p-2}\int P_{n,p}(\lambda_2)\left[r\frac{\partial f}{\partial r} - nf(x)\right]\sigma_r^a(dx) \tag{2.14}$$

$$(2-p)\,\alpha = r^{p-1}\int \frac{\partial f}{\partial r}\,\sigma_r^a(dx), \tag{2.15}$$

and, as usual, the $2-p$ on the left is replaced by -1 when $p = 2$. Also

$$K = \begin{cases} \int \left(f(x) + \dfrac{R}{p-2} \dfrac{\partial f}{\partial R} \right) \sigma_R^a(dx), \; p \geqslant 3 \\ \int \left(f(x) - \log R \dfrac{\partial f}{\partial R} \right) \sigma_R^a(dx), \; p = 2 \end{cases} \tag{2.16}$$

Since $P_{n,p}(\lambda)$ is a polynomial in λ the integrals above are linear combinations of terms of the type

$$\int \prod_{\lambda=1}^{p} \left[\frac{(x_i - a_i)(y_i - a_i)}{|x-a||y-a|} \right]^{r_i} \{ \ldots \} \, \sigma^a(dx)$$

and this last integral is a constant multiple of

$$\prod_{\lambda=1}^{p} \left(\frac{y_i - a_i}{|y-a|} \right)^{r_i}$$

the constant depending on a, R, r. Consequently Y_n and Z_n are polynomial functions of $u_1, \ldots, u_p$ where $u = (u_1, \ldots, u_p) = (y-a)/|y-a|$ and so are independent of $|y-a|$.

Thus, summing up, we have

THEOREM 2.42. *Suppose that $R > r > 0$, that $A = B_R(a) \setminus \bar{B}_r(a)$ and that f is harmonic in an open set containing $\bar{A}$. Then, for $y \in A$,*

$$f(y) = K + \alpha h(y) + \sum_{n=1}^{\infty} Y_n(u) \, \rho^n + \sum_{n=1}^{\infty} Z_n(u) \, \rho^{2-p-n} \tag{2.17}$$

where $\rho = |y-a|$, $u = (y-a)/|y-a|$ and *K, α, Y_n and Z_n are given by* (2.13), (2.14), (2.15) *and* (2.16).

The polynomials Y_n and Z_n are said to be spherical harmonics. We have seen, in Chapter 1, §6, that $\rho^n P_{n,p}(\lambda)$ are harmonic in y and so, from the form of (2.13) and (2.14) it follows that $\rho^n Y_n(u)$ and $\rho^n Z_n(u)$ are harmonic in y. It is this that enables us to give an elegant proof of the following orthogonality relations.

LEMMA 2.43.

$$\mathscr{S}_1^0(Y_n) = \mathscr{S}_1^0(Z_n) = 0; \; \mathscr{S}_1^0(Y_n Y_m) = \mathscr{S}_1^0(Y_n Z_m) = \mathscr{S}_1^0(Z_n Z_m) = 0$$

for $m \neq n$. Also $\mathscr{S}_1^0(Y_n Z_n) = 0$.

Let $Q_n = \rho^n Y_n(u)$. Then, by Lemma 2.12, with a the zero vector, $r = 0$ and $R = 1$ and noting that Q_n and Q_m are harmonic, we have

$$\int \left(Q_n \frac{\partial Q_m}{\partial \rho} - Q_m \frac{\partial Q_n}{\partial \rho} \right) \sigma_1^0(dy) = 0.$$

Since $\dfrac{\partial Q_n}{\partial \rho} = nY_n(u)$ when $\rho = 1$, this gives, since $m \neq n$, $\mathscr{S}_1^0(Y_n Y_m) = 0$.

A similar proof shows that $\mathscr{S}_1^0(Y_n Z_m) = \mathscr{S}_1^0(Z_n Z_m) = \mathscr{S}_1^0(Y_n Z_n) = 0$ and, using $\rho^n Y_n$ and 1, and $\rho^n Z_n$ and 1, in turn in Lemma 2.12 as above we have

$$\mathscr{S}_1^0(Y_n) = \mathscr{S}_1^0(Z_n) = 0.$$

The expansion in Theorem 2.42 is clearly the analogue for harmonic functions of the Laurent Expansion for analytic functions, and, like the latter, it is unique. To show this, we note that K, α, Y_n and Z_n are linear functionals of f and thus we need only show that if $f = 0$ then all coefficients must be zero.

Since the expansions for $h_y(x)$ are, for any ρ strictly between r and R, uniformly convergent in λ the expansion for f converges uniformly with respect to u. Setting $f = 0$, multiplying through by Y_n and integrating with respect to σ_ρ^a we obtain

$$\rho^n \mathscr{S}_1^0(Y_n^2) = 0.$$

because of the relations in Lemma 2.43. Consequently $Y_n = 0$. A similar argument shows that $Z_n(u) = 0$. Finally, we integrate the expansion as it stands to obtain

$$K + \alpha h_a(y) = 0$$

with $|y - a| = \rho$ and so $K = \alpha = 0$.

Suppose now that f is harmonic in an open set containing $\bar{B}_R(a) \setminus (a)$. Then, for $y \in B^* = B_R(a) \setminus (a)$, by (2.17),

$$f(y) = g(y) + \alpha h_a(y) + \sum_{n=1}^{\infty} Z_n(u)\, \rho^{2-p-n}$$

where

$$g(y) = K + \sum_{n=1}^{\infty} Y_n(u)\, \rho^n.$$

The right-hand side of this last is harmonic in $B_R(a)$, and so we may define $g(a) = K$ and thus extend g, initially defined in B^*, to a function harmonic in $B_R(a)$. In fact, g has a **removable** singularity at a.

The nature of the singularity of f at a is governed by

THEOREM 2.44. *Suppose f is harmonic near but not at a. Then (a) A necessary and sufficient condition that f have a pole of order m at a, that is, that $Z_n(u)$ is identically zero for $n > m$ and $Z_m(u)$ is not identically zero, is that*

$$\limsup_{\rho \to 0} \rho^{m+p-2} \mathscr{S}_\rho^a(f_+) < +\infty, \tag{2.18}$$

and

$$\limsup_{\rho\to 0} \rho^{m+p-2}\,\mathscr{S}^a_\rho(f_+) > 0. \tag{2.19}$$

(*b*) *A necessary and sufficient condition that f have a removable singularity at a is that*

$$\lim_{\rho\to 0} (h_a(y))^{-1}\,\mathscr{S}^a_\rho(|f|) = 0. \tag{2.20}$$

First, by Lemma 2.43,

$$\mathscr{S}^a_\rho(f) = K+\alpha h_a(y) \tag{2.21}$$

and so $\lim_{\rho\to 0} \rho^{m+p-2}\,\mathscr{S}^a_\rho(f) = 0$ for $m \geqslant 1$. Since $|f| = 2f_+ - f$ it follows from (2.18) that

$$\limsup_{\rho\to 0} \rho^{m+p-2}\,\mathscr{S}^a_\rho(|f|) < +\infty \tag{2.18'}$$

and from (2.19) that

$$\liminf_{\rho\to 0} \rho^{m+p-2}\,\mathscr{S}^a_\rho(|f|) > 0. \tag{2.19'}$$

Now

$$\rho^{m+p-2}\,\mathscr{S}^a_\rho(fZ_n) = \rho^{m-n}\,S^0_1(Z^2_n)$$

From (2.18′), this requires, for $n > m$, that $\mathscr{S}^0_1(Z^2_n) = 0$ and we thus have $Z_n(u)$ identically zero.

If, also, $Z_m(u)$ is identically zero then, from (2.17), $|f| = o(\rho^{2-p-m})$ which contradicts (2.19′). So Z_m is not identically zero. This proves the sufficiency of the conditions.

Conversely, if f has a pole of order m at a then, by (2.17), and as $\rho \to 0$,

$$\rho^{m+p-2}\,\mathscr{S}^a_\rho(|f|) = \mathscr{S}^a_\rho(|Z_m|)+0(1)$$

which gives (2.18′). Since $Z_m(u)$ is continuous and not identically zero there is a $k > 0$ and a subset M of $S_1(0)$ with $\sigma^0_1(M) \neq 0$ in which $|Z_m(u)| > k$. Consequently $S^\rho_a(|Z_m|) > 0$ which now gives (2.19′). This, with (2.21) gives (2.18) and (2.19).

(*b*) Suppose (2.20) holds. We have $\mathscr{S}^a_\rho(|f|) \geqslant |\mathscr{S}^a_\rho(f)|$ so, by (2.21),

$$|K(h_a(y))^{-1}+\alpha| \to 0 \text{ as } \rho \to 0$$

so that $\alpha = 0$. Also, if (2.20) holds, so does (2.18) for $m = 0$ when $p \geqslant 3$ and for $m = \frac{1}{2}$ when $p = 2$; and so $Z_n(u)$ is identically zero for $n > 0$. Consequently $f(y) = g(y)$, and so has a removable singularity at a.

Conversely, if f has a removable singularity at a then f is bounded near a, and then (2.20) follows.

We may note that, since $-f$ is harmonic with f, and Y_n, Z_n are linear functionals of f, we may in (2.18) and (2.19) replace f_+ by f_-.

§ 2.8. Behaviour at ω

Suppose that f is harmonic in an open set containing $\mathbb{R}^p \setminus B_r(a)$. Then, for $y \in \mathbb{R}^p \setminus \bar{B}_r(a)$,

$$f(y) = K + \alpha h_a(y) + \sum_{n=1}^{\infty} Y_n(u)\, \rho^n + \sum_{n=1}^{\infty} Z_n(u)\, \rho^{2-p-n}$$

and the behaviour in the neighbourhood of ω is determined by

THEOREM 2.45. *Let f be harmonic in a neighbourhood of ω. Then:—a necessary and sufficient condition that f have polynomial growth of order m at ω, that is, that $Y_n(u)$ be identically zero for $n > m$ and $Y_m(u)$ be not identically zero, is that*

$$0 < \limsup_{\rho \to \infty} \rho^{-m}\, \mathscr{S}^a_\rho(f_+) < +\infty. \tag{2.23}$$

We have

$$\rho^{-m}\, \mathscr{S}^a_\rho(fY_n) = \rho^{n-m}\, \mathscr{S}^0_1(Y_n^2)$$

and, arguing as in the proof of Theorem 2.44(a) we show that the left-hand side remains bounded as $\rho \to \infty$. This then requires, for $n > m$, that $\mathscr{S}^0_1(Y_n^2) = 0$, and so $Y_n(u)$ is identically zero. Then

$$|f(y)| \leqslant |g^*(y)| + |\alpha|\, |h_a(y)| + \sum_{n=1}^{m} \rho^n\, |Y_n(u)|, \tag{2.24}$$

where

$$g^*(y) = K + \sum_{n=1}^{\infty} \rho^{2-p-n}\, z_n(u).$$

If $Y_m(u)$ were identically zero, then $f(y) = 0(\rho^{m-1}) = 0(\rho^m)$ which would contradict the left-hand inequality in (2.23).

To prove the necessity, we proceed as in the proof of Theorem 2.44(a).

Arguing in much the same way, we may show that a necessary and sufficient condition that $Y_n(u)$ be identically zero, for $n > 0$, is that

$$\limsup_{\rho \to \infty} \mathscr{S}^a_\rho(f_+) < +\infty.$$

There is an important corollary of this last theorem—an analogue of Picard's Theorem for analytic functions—which we state as

THEOREM 2.46. *Suppose that f is harmonic in $\mathbb{R}^p$. Then a necessary and sufficient condition that f be a polynomial exactly of degree m is that*

$$\mathscr{S}^a_\rho(f_+) = 0(\rho^m) \text{ as } \rho \to \infty. \tag{2.25}$$

Since f is harmonic in $\mathbb{R}^p$, it has no singularity at a, and so $g^*(y) = K$. Theorem 2.45 then gives the result.

In particular, we note that, if a function harmonic in $\mathbb{R}^p$ is bounded above, then it is a constant.

As before, we remark that in (2.23) and (2.25) f_+ may be replaced by f_-. Furthermore, **if (2.23) (and (2.25)) holds for any given $a \in \mathbb{R}^p$, it holds for all $a \in \mathbb{R}^p$.** For if (2.23) holds, then in the Laurent expansion about a, $Y_n(u)$ is identically zero for $n > m$. If $b \in \mathbb{R}^p$, $b \neq a$ then $\rho^n Y_n(u)$, being a polynomial of degree n in $y-a$, is a harmonic polynomial of degree n in $y-b$. Hence the Laurent expansion of f about b will have zero coefficients for $|y-b|^n$ when $n > m$. From this it follows that (2.23) holds with b in the place of a.

It may have been a source of irritation to the reader that, having defined harmonicity at ω, we have so far only one example of a function which is harmonic there, namely, the constant function. We are now in a position to remedy this state of affairs.

Suppose that f is harmonic in the neighbourhood of ω and has the Laurent expansion about a of the form

$$f(y) = K + \sum_{n=1}^{\infty} \rho^{2-p-n} Z_n(u)$$

valid for $\rho = |y-a| > r$, say. Then $f(y)-K = O(\rho^{1-p})$ as $\rho \to \infty$; and so, for $b \in \mathbb{R}^p$, $f(y)-K = O(|y-b|^{1-p})$. Consequently, the Laurent expansion of f about b must be of the form

$$f(y) = K + \sum_{n=1}^{\infty} (\rho')^{2-p-n} Z_n'(u),$$

and hence $\mathscr{S}_\rho^b(f) = K$. Since also $\lim_{y\to\omega} f(y) = K$ we have, if we set $f(\omega) = K$, that f is harmonic at ω.

We saw earlier that the Poisson Integral solves the Dirichlet Problem both for $B_R(a)$ and $\mathbb{R}^p \setminus \bar{B}_R(a)$. We have seen, too, that for $B_R(a)$ this solution is unique. But it is not unique for $\mathbb{R}^p \setminus \bar{B}_R(a)$, since, for any constant K,

$$g(x) = I_R^a\,(f-K)\,(x) + K$$

is harmonic in $\mathbb{R}^p \setminus \bar{B}_R(a)$ and $\lim_{x\to\xi} g(x) = f(\xi) - K + K = f(\xi)$, $\xi \in S_R(a)$.

It is instructive to analyse what is happening here. Clearly, given the continuous data on $S_R(a)$, it is still open to us to specify what must happen at ω, and here, indeed, we are saying that $\lim_{x\to\omega} f(x) = K$. Thus, rightly, ω should be regarded as part of the frontier of the region. Alternatively, we could seek a function harmonic in $\overline{\mathbb{R}}^p \setminus \bar{B}_R(a)$, i.e. is harmonic at ω also, which matches up with f on $S_R(a) = \partial\,[\overline{\mathbb{R}}^p \setminus \bar{B}_R(a)]$.

We present a modified Poisson Integral, due to Brelot, which does just this. Let, for $|x-a| > R$,

$$L_R^a(x, t) = K_R^a(x, t) - R^{p-2} |x-a|^{2-p} + 1$$

and define the modified Poisson Integral of f, $J_R^a f$, by

$$(J_R^a f)(x) = \int f(t) L_R^a(x, t) \sigma_R^a(dt).$$

Then

$$(J_R^a f)(x) = (I_R^a f)(x) + \int (1 - R^{p-2} |x-a|^{2-p}) f(t) \sigma_R^a(dt)$$

and thus, for $\xi \in S_R(a)$,

$$\lim_{x \to \xi} (J_R^a f)(x) = \lim_{x \to \xi} (I_R^a f)(x) = f(\xi).$$

By (1.17) and (1.18) (or (1.18′)), for $|x-a| > |t-a|$,

$$L_R^a(x, t) = 1 + \sum_{n=1}^{\infty} \frac{2n+p-2}{p-2} \frac{R^{n+p-2}}{|x-a|^{n+p-2}} P_{n,p}(\lambda)$$

and so L_R^a is, qua function of x, harmonic at ω. Consequently

$$\mathscr{S}_\rho^a(L_R^a) = L_R^a(\omega, t) = 1,$$

and thus

$$\mathscr{S}_\rho^b(J_R^a f) = \int f(t) \mathscr{S}_\rho^b(L_R^a) \sigma_R^a(dt) = \int f(t) \sigma_R^a(dt)$$

while

$$(J_R^a f)(\omega) = \int f(t) \sigma_R^a(dt).$$

Thus $J_R^a f$ is the function we are seeking. Furthermore, if u is harmonic in an open set $G \supset \overline{\mathbb{R}}^p \setminus \bar{B}_R(a)$, then it coincides with $J_R^a f$ in $\overline{\mathbb{R}}^p \setminus \bar{B}_R(a)$ where $f = \operatorname{rest}_{S_R(a)} u$. This follows as before from Theorem 2.17.

Just as we were able to deduce the Harnack Inequalities from the Poisson Integral, so we can deduce Harnack Inequalities pivoted on ω from the extended Poisson Integral. Thus, if f is harmonic and non-negative in an open set G containing ω, then for R such that $\overline{\mathbb{R}}^p \setminus B_R(a) \subset G$ we have, when $x \in \overline{\mathbb{R}}^p \setminus \bar{B}_R(a)$,

$$f(\omega) R^{p-2} \left\{ \frac{|x-a|-R}{(|x-a|+R)^{p-1}} - \frac{1}{|x-a|^{p-2}} + \frac{1}{R^{p-2}} \right\} \leqslant f(x)$$

$$\leqslant f(\omega) R^{p-2} \left\{ \frac{|x-a|+R}{(|x-a|-R)^{p-1}} - \frac{1}{|x-a|^{p-2}} + \frac{1}{R^{p-2}} \right\},$$

and from this we may deduce the following variant of Theorem 2.20.

THEOREM 2.47. *If u is harmonic and non-negative in a domain $G \subset \overline{\mathbb{R}}^p$ then, given any closed set $F \subset G$, there are positive constants α and β depending only on F, such that*

$$\alpha < u(x)/u(y) < \beta \text{ for } x, y \in F$$

Furthermore, only small variations in the proofs are required to demonstrate that Theorems 2.22, 2.23, 2.23.1, 2.24 and 2.25 also hold in $\overline{\mathbb{R}}^p$.

Chapter 3

THE CONDUCTOR PROBLEM AND CAPACITY

§ 3.1. The Riesz Composition Formula

THEOREM 3.1. *Suppose that* $0 < \alpha < p$, $0 < \beta < p$ *and* $0 < \alpha+\beta < p$. *Then*

$$\int_{\mathbb{R}^p} | x-z |^{\alpha-p} | z-y |^{\beta-p} \, dz = k_{\alpha,\beta} | x-y |^{\alpha+\beta-p}$$

where

$$k_{\alpha,\beta} = \pi^{\frac{1}{2}p} \frac{\Gamma(\frac{1}{2}\alpha)\,\Gamma(\frac{1}{2}\beta)\,\Gamma((p-\alpha-\beta)/2)}{\Gamma\left(\frac{p-\alpha}{2}\right)\Gamma\left(\frac{p-\beta}{2}\right)\Gamma(\frac{1}{2}(\alpha+\beta))}$$

This is the famous **Riesz Composition Formula.**

First—and there is no great difficulty about this—we prove that the formula holds with some constant. Make the transformation

$$z = y+| x-y | t$$

in the integral on the left-hand side to obtain, with $w = (x-y)/| x-y |$,

$$| x-y |^{\alpha+\beta-p} \int_{\mathbb{R}^p} | w-t |^{\alpha-p} | t |^{\beta-p} \, dt.$$

This last integral is, in fact, not dependent on the unit vector w. For suppose that w_1 and w_2 are two unit vectors, and let U be an orthogonal matrix such that $w_2 = Uw_1$. Then, making the substitution $t = Us$ we obtain

$$\int_{\mathbb{R}^p} | w_2-t |^{\alpha-p} | t |^{\beta-p} \, dt = \int_{\mathbb{R}^p} | U(w_1-s) |^{\alpha-p} | Us |^{\beta-p} | \det U | \, dt$$

which, since U is orthogonal, reduces to

$$\int_{\mathbb{R}^m} | w_1-s |^{\alpha-p} | s |^{\beta-p} \, ds.$$

To evaluate the constant is a more difficult matter, and, following a treatment due to Deny, I present the argument by way of a number of lemmas.

LEMMA 1. *Let* $x \, . \, y = x_1 y_1 + \ldots + x_p y_p$. *Then*

$$\int_{\mathbb{R}^p} | y |^{\alpha - p} e^{-2i\pi x . y} \, dy = C_\alpha | x |^{-\alpha}$$

where C_α *is a constant depending on* α.

Suppose that U is an orthogonal matrix of which the first column is $x/| x |$. Consider the transformation $y = Uz$. Since $U^{-1} = U^T$ we have $| x | z_1 = x . y$, and so the integral becomes

$$\int_{\mathbb{R}^p} | z |^{\alpha - p} e^{-2i\pi | x | z_1} \, dz.$$

In this latter integral we set $z = t/| x |$ and obtain

$$| x |^{-\alpha} \int_{\mathbb{R}^p} | t |^{\alpha - p} e^{-2i\pi t_1} \, dt.$$

LEMMA 2.

$$\int_{\mathbb{R}^p} e^{-\pi | y |^2} e^{-2i\pi x . y} \, dy = e^{-\pi | x |^2}.$$

If we expand $\pi | y |^2 + 2i\pi \, x \, . \, y$ and transform the integral to a repeated integral we obtain a product of p integrals of the form

$$\int_{-\infty}^{\infty} e^{-\pi(\eta^2 + 2i\xi\eta)} \, d\eta$$

which may be rewritten as

$$e^{-\pi\xi^2} \int_{-\infty}^{\infty} e^{-\pi(\eta + i\xi)^2} \, d\eta.$$

Since $e^{-\pi z^2}$ is analytic, an integration around a rectangular contour shows this last integral is equal $\int e^{-\pi\eta^2} \, d\eta$ which has value 1. The lemma now follows.

LEMMA 3.

$$\int_{\mathbb{R}^p} e^{-\pi | x |^2} | x |^{\alpha - p} \, dx = C_\alpha \int_{\mathbb{R}^p} e^{-\pi | x |^2} | x |^{-\alpha} \, dx$$

where C_α *is the constant occurring in Lemma* 1.

By Lemma 1 the right-hand side may be written in the form

$$\int_{\mathbb{R}^p} \int_{\mathbb{R}^p} e^{-\pi | x |^2} | y |^{\alpha - p} e^{-2\pi i x . y} \, dy \, dx.$$

Changing the order of integration, and using Lemma 2, this becomes

$$\int_{\mathbb{R}^p} e^{-\pi | y |^2} | y |^{\alpha - p} \, dy,$$

giving the lemma.

In Lemma 3 we transform to spherical polar coordinates in each integral, to obtain

$$\int_0^\infty e^{-\pi r^2} r^{\alpha-1}\, dr = C_\alpha \int_0^\infty e^{-\pi r^2} r^{p-\alpha-1}\, dr.$$

Setting $\pi r^2 = t$ we then have

$$\pi^{-\frac{1}{2}\alpha}\, \Gamma(\tfrac{1}{2}\alpha) = C_\alpha\, \pi^{-\frac{1}{2}(p-\alpha)}\, \Gamma(\tfrac{1}{2}(p-\alpha))$$

so that

$$C_\alpha = \pi^{\frac{1}{2}p-\alpha}\, \Gamma(\tfrac{1}{2}\alpha)/\Gamma(\tfrac{1}{2}(p-\alpha)). \tag{3.1}$$

LEMMA 4.

$$\int_{\mathbb{R}^p} \int_{\mathbb{R}^p} |x|^{\alpha-p}\, |y-x|^{\beta-p}\, e^{-2i\pi y.z}\, dx\, dy = C_\alpha C_\beta\, |z|^{-\alpha-\beta}$$

Changing the order of integration, the integral becomes

$$\int_{\mathbb{R}^p} |x|^{\alpha-p} \int_{\mathbb{R}^p} |y-x|^{\beta-p}\, e^{-2i\pi y.z}\, dy\, dx.$$

In the inner integral, set $y-x = t$ to obtain

$$\Big(\int_{\mathbb{R}^p} |x|^{\alpha-p}\, e^{-2i\pi x.z}\, dx\Big)\, \Big(\int_{\mathbb{R}^p} |t|^{\beta-p}\, e^{-2i\pi t.z}\, dt\Big)$$

which, by Lemma 1, becomes $C_\alpha\, |z|^{-\alpha}\, C_\beta\, |z|^{-\beta}$.

Finally,

$$\int_{\mathbb{R}^p} |x|^{\alpha-p}\, |y-x|^{\beta-p}\, dx = k_{\alpha,\beta}\, |y|^{\alpha+\beta-p}.$$

Multiplying through by $e^{-2i\pi y.z}$ and integrating with respect to y, we have, using Lemma 1 and Lemma 4,

$$C_\alpha C_\beta\, |z|^{-\alpha-\beta} = k_{\alpha,\beta}\, C_{\alpha+\beta}\, |z|^{-\alpha-\beta}$$

so that

$$k_{\alpha,\beta} = C_\alpha C_\beta / C_{\alpha+\beta}\,.$$

This now gives the Riesz Composition Formula.

§ 3.2. The Riesz Fractional Integral

The *Riesz kernel* $k_\alpha(x)$ is defined, for $0 < \alpha < p$, by

$$k_\alpha(x) = \pi^{\alpha-\frac{1}{2}p}\, \frac{\Gamma(\frac{1}{2}(p-\alpha))}{\Gamma(\frac{1}{2}\alpha)}\, |x|^{\alpha-p}.$$

The Riesz Composition Formula may be then expressed in the form

$$(k_\alpha * k_\beta)\,(x) = k_{\alpha+\beta}\,(x) \tag{3.2}$$

where $*$ denotes convolution.

Suppose now that f is a test function. Then, for $0 < \alpha < p$, we define the Riesz fractional integral U_α^f of f by

$$U_\alpha^f(x) = (k_\alpha * f)(x) \tag{3.3}$$

where $*$ denotes convolution. It then follows that, for $0 < \alpha < p$, $0 < \beta < p, 0 < \alpha+\beta < p$, that

$$\begin{aligned} U_{\alpha+\beta}^f(x) &= (k_{\alpha+\beta} * f)(x) = [(k_\alpha * k_\beta) * f](x) \\ &= [k_\alpha * (k_\beta * f)](x) = (k_\alpha * U_\beta^f)(x) \end{aligned}$$

by changing the order of integration in the triple integrals involved. This result says, in effect, that the $(\alpha+\beta)$-th integral of f is the same as the α-th integral of the β-th integral of f.

Now

$$U_\alpha^f(x) = \pi^{\alpha-\frac{1}{2}p} \frac{\Gamma(\frac{1}{2}(p-\alpha))}{\Gamma(\frac{1}{2}\alpha)} \int_{\mathbb{R}^p} |x-y|^{\alpha-p} f(y)\, dy$$

and so U_α^f is defined for all complex α for which $re(\alpha) > 0$ other than the values $p+2n$ ($n = 0, 1, 2, \ldots$) and, in this region of the complex plane, is an analytic function of α.†

We proceed now to continue U_α^f analytically into the half-plane $re(\alpha) \leqslant 0$. In this process the factor outside the integral is of crucial importance and this mainly because $\Gamma(\frac{1}{2}\alpha)$ has poles at $\alpha = -2n$ ($n = 0, 1, 2, \ldots$).

Consider the expression $J^{(m)}f$ defined, for any positive integer m, by

$$(J^m f)(x) = T_1^{(m)} + T_2 + T_3^{(m)}$$

where

$$T_1^{(m)} = \pi^{\alpha-\frac{1}{2}p} \frac{\Gamma(\frac{1}{2}(p-\alpha))}{\Gamma(\frac{1}{2}\alpha)} \int_{B_1} |y|^{\alpha-p} (f(x+y) - P_m(f; x, |y|))\, dy$$

$$T_2 = \pi^{\alpha-\frac{1}{2}p} \frac{\Gamma(\frac{1}{2}(p-\alpha))}{\Gamma(\frac{1}{2}\alpha)} \int_{\mathbb{R}^p \setminus B_1} |y|^{\alpha-p} f(x+y)\, dy$$

$$T_3^{(m)} = \frac{2\pi^\alpha}{\Gamma(\frac{1}{2}p)} \frac{\Gamma(\frac{1}{2}(p-\alpha))}{\Gamma(\frac{1}{2}\alpha)} \sum_{s=0}^{m} h_s(\Delta^s f)(x) \frac{1}{2s+\alpha}$$

where B_1 is the unit ball centred at 0, $h_s = (2^s s!\, p(p+2)\ldots(p+2s-2))^{-1}$ and

$$P_m(f; a, r) = \sum_{s=0}^{m} h_s(\Delta^s f)(a)\, r^{2s}.$$

† We define ρ^α to be $e^{\alpha \log \rho}$ and this is then quite unambiguous.

The reader will no doubt recall that $P_m(f; a, r)$ is the sum of the first $(m+1)$ terms in Pizzetti's formula in Chapter 1, §7. Since this is so

$$\mathscr{S}^x_{|y|}(f)-P_m(f; x, |y|) = 0(|y|^{2m+2})$$

and so the integral in $T_1^{(m)}$ exists, and is an analytic function of α for $re(\alpha) > -2(m+1)$, and so therefore is $T_1^{(m)}$, except at the values $p+2n$ $(n = 0, 1, 2 \ldots)$.

The integral in T_2 exists, and is analytic in α for all α (recall that f, being a test function, is of compact support) and so T_2 is analytic everywhere other than at $(p+2n)$.

Since

$$[\Gamma(\tfrac{1}{2}\alpha)(2s+\alpha)]^{-1} = \tfrac{1}{4}\alpha(1+\tfrac{1}{2}\alpha)\ldots(s-1+\tfrac{1}{2}\alpha)(\Gamma(s+1+\tfrac{1}{2}\alpha))^{-1}$$

it is analytic everywhere and so $T_3^{(m)}$ is analytic in α except at $p+2n$.

Thus $J^{(m)}f$ is analytic in α in the half-plane $re(\alpha) > -2(m+1)$ apart from the points $p+2n$.

Now

$$T_1^{(m)} = T_1^{(m+1)}+\frac{\pi^{\alpha-\frac{1}{2}p}\,\Gamma(\frac{1}{2}(p-\alpha))}{\Gamma(\frac{1}{2}\alpha)} \int_{B_1} h_{m+1}(\Delta^{m+1}f)(x)\,|y|^{\alpha+2m-p+2}\,dy$$

and

$$\int_{B_1}|y|^{\alpha+2m-p+2}\,dy = \frac{2\pi^{\frac{1}{2}p}}{\Gamma(\frac{1}{2}p)}\int_0^1 r^{\alpha+2m+1}\,dr,$$

so that

$$\begin{aligned} T_1^{(m)} &= T_1^{(m+1)}+\frac{2\pi^{\alpha}}{\Gamma(\frac{1}{2}p)}\frac{\Gamma(\frac{1}{2}(p-\alpha))}{\Gamma(\frac{1}{2}\alpha)}\,h_{m+1}(\Delta^{m+1}f)(x)(\alpha+2m+2)^{-1} \\ &= T_1^{(m+1)}+T_3^{(m+1)}-T_3^{(m)} \end{aligned}$$

so that, for $re(\alpha) > -2(m+1)$ and $\alpha \neq (p+2n)$

$$T_1^{(m)}+T_3^{(m)} = T_1^{(m+1)}+T_3^{(m+1)}$$

and hence

$$(J^{(m)}f)(x) = (J^{(m+1)}f)(x) \text{ for these } \alpha.$$

Furthermore, $J^{(m+1)}f$ is analytic in α for $re(\alpha) > -2(m+2)$, $\alpha \neq (p+2n)$. It is therefore the analytic continuation of $J^{(m)}f$. In particular,

$$\begin{aligned} (J^0f)(x) &= T_1^0+T_2+T_3^{(0)} \\ &= \pi^{\alpha-\frac{1}{2}p}\frac{\Gamma(\frac{1}{2}(p-\alpha))}{\Gamma(\frac{1}{2}\alpha)}\left\{\int_{B_1}|y|^{\alpha-p}(f(x+y)-f(x))\,dy\right. \\ &\qquad \left.+\int_{\mathbb{R}^p\setminus B_1}|y|^{\alpha-p}f(x+y)\,dy+\frac{2\pi^{\frac{1}{2}p}}{\Gamma(\frac{1}{2}p)}\frac{f(x)}{\alpha}\right\} \end{aligned}$$

for $re(\alpha) > -2$.

Also, for $re(\alpha) > 0$, $\int_{B_1} |y|^{\alpha-p}\, dy = \dfrac{2\pi^{\frac{1}{2}p}}{\Gamma(\frac{1}{2}p)} \dfrac{1}{\alpha}$ so that $(J^0 f)(x) = U^f_\alpha(x)$ for $re(\alpha) > 0$.

Thus $J^{(m)}f$ is the analytic continuation of the Riesz fractional integral, initially defined for $re(\alpha) > 0$, into the larger region $re(\alpha) > -2(m+1)$. Thus, the sequence $\{J^{(m)}f\}$ defines the analytic continuation through the whole of $re(\alpha) \leqslant 0$. By an abuse of notation, we shall denote this analytic function by U^f_α. It is clear that $U^f_\alpha(x)$ is an infinitely differentiable function of x.

When $\alpha = 2m$, $T_1^{(m)} = T_2 = 0$ and in $T_3^{(m)}$ all terms except the last vanish. The last term has the value

$$\frac{2\pi^{-2m}}{\Gamma(\frac{1}{2}p)} \Gamma(\tfrac{1}{2}p+m)\, h_m(\Delta^m f)(x) \lim_{\alpha\to -2m} (\Gamma(\tfrac{1}{2}\alpha)\, 2m+\alpha)^{-1}.$$

Since

$$[\Gamma(\tfrac{1}{2}\alpha)\,(2m+\alpha)]^{-1} = \frac{\frac{1}{2}\alpha(1+\frac{1}{2}\alpha)\ldots(m-1+\frac{1}{2}\alpha)}{2\Gamma(\frac{1}{2}\alpha+m+1)},$$

we have

$$U^f_{-2m}(x) = \frac{\pi^{-2m}\,\Gamma(\frac{1}{2}p+m)\,(-m)\,(1-m)\ldots(-1)}{\Gamma(\frac{1}{2}p)\, 2\Gamma(1)\, 2^m m!\, p(p+2)\ldots(p+2m-2)} (\Delta^m f)(x)$$

$$= \left(-\frac{1}{4\pi^2}\right)^m (\Delta^m f)(x),$$

and so the Riesz fractional integral of order $-2m$ inverts the polyharmonic operator Δ^m. In particular

$$U^f_0(x) = f(x).$$

Finally, for $0 < \alpha < \frac{1}{2}p$, $0 < \beta < \frac{1}{2}p$,

$$(k_\alpha * U^f_\beta)(x) = U^f_{\alpha+\beta}(x),$$

and so, by analytic continuation, this holds for all complex α, β (other than $p+2n$). In particular,

$$(k_\alpha * U^f_{-\alpha})(x) = U^f_0(x) = f(x).$$

This gives a result we shall find very useful later.

LEMMA 3.2. *Given a test function f, and $0 < \alpha < p$, we can find an infinitely differentiable function ϕ such that*

$$f(x) = (k_\alpha * \phi)(x).$$

§ 3.3. The Riesz Potentials

Let $\mathfrak{B}$ be the Borel tribe in $\mathbb{R}^p$. A mapping

$$\gamma : \mathfrak{B} \to \mathbb{R}$$

such that

(*i*) *γ is completely additive on $\mathfrak{B}$.*

(*ii*) *$\gamma(K)$ is finite for each compact K.*

(*iii*) *γ is regular,* that is, given $B \in \mathfrak{B}$ and $\varepsilon > 0$ there is a closed set $F \subset B$ and an open set $G \supset B$ such that, when $E \subset G \setminus F$, $|\gamma(E)| < \varepsilon$ is said to be a **charge** on $\mathbb{R}^p$.

Clearly, if μ_1 and μ_2 are Radon measures on $\mathbb{R}^p$ then $\mu_1 - \mu_2$ is a charge provided that either μ_1 or μ_2 is real-valued. Conversely, any charge γ is the difference of two such measures. For, if we set

$$\gamma^+(B) = \sup \{\gamma(E) \mid E \subset B, E \in \mathfrak{B}\}$$
$$\gamma^-(B) = -\inf \{\gamma(E) \mid E \subset B, E \in \mathfrak{B}\}$$

then it may be shown that γ^+ and γ^- are measures and that $\gamma = \gamma^+ - \gamma^-$. γ^+ and γ^- are known, respectively, as the positive and negative variations of γ. Furthermore, if $\gamma = \mu_1 - \mu_2$ then $\mu_1 \geqslant \gamma^+$ and $\mu_2 \geqslant \gamma^-$.

Suppose that f is measurable and integrable with respect to each of γ^+ and γ^-. Then we define the integral of f with respect to γ by

$$\int f(x)\, \gamma(dx) = \int f(x)\, \gamma^+(dx) - \int f(x)\, \gamma^-(dx).$$

We shall suppose from now on that $0 < \alpha < p$. The α-potential of a charge γ, U_α^γ, is defined by

$$U_\alpha^\gamma(x) = \int k_\alpha(x-y)\, \gamma(dy).$$

$U_\alpha^\gamma(x)$ will, in fact, be defined if and only if not both of $U_\alpha^{\gamma^+}(x)$ and $U_\alpha^{\gamma^-}(x)$ are infinite. By contrast, the α-potential of a measure μ is always defined, finite or infinite.

Lemma 3.3. *If the measure μ is of compact support then $U_\alpha^\mu(x)$ is finite a.e.*

We can find $r_1 > 0$ such that $\operatorname{supp} \mu \subset B_{r_1}(0)$. Let $r > 0$. Then

$$\int_{B_r} U_\alpha^\mu(x)\, dx = \int_{B_r} \int k_\alpha(x-y)\, \mu(dy)\, dx,$$

and this last is a constant multiple of

$$\int_{B_{r_1}} \int_{B_r} |x-y|^{\alpha-p}\, dx\, \mu(dy).$$

Now, for $|y| < r_1$,

$$\int_{B_r} |x-y|^{\alpha-p}\,dx < \int_{B_{r+r_1}} |x|^{\alpha-p}\,dx = K, \text{ say;}$$

and hence

$$\int_{B_r} U_\alpha^\mu(x)\,dx < K' \int_{B_{r_1}} \mu(dy) = K'\,\mu\,(\mathbb{R}^p).$$

Thus U_α^μ is integrable in $B_r(0)$ and so is finite a.e. there. Since r was arbitrary, the lemma now follows.

There is an all-or-none principle analogous to that for superharmonic functions which operates also for α-potentials. We have, in fact,

THEOREM 3.4. *A necessary and sufficient condition that the α-potential of a measure μ be finite a.e. in $\mathbb{R}^p$ is that*

$$\int_{|y|\geqslant 1} |y|^{\alpha-p}\,\mu(dy) < +\infty. \tag{3.4}$$

Furthermore, if (3.4) does not hold then $U_\alpha^\mu(x)$ is identically infinite in $\mathbb{R}^p$.

First, for $|y| > |x|+1$,

$$\frac{|x-y|}{|y|} \leqslant \frac{|x|+|y|}{|y|} \leqslant \frac{2|x|+1}{|x|+1}$$

and so a strictly positive constant multiple of $U_\alpha^\mu(x)$ is not less than

$$\int_{|y|>|x|+1} |x-y|^{\alpha-p}\,\mu\,(dy) \geqslant \left(\frac{2|x|+1}{|x|+1}\right)^{\alpha-p} \int_{|y|>|x|+1} |y|^{\alpha-p}\,\mu\,(dy).$$

When $r > 1$,

$$\int_{|y|\geqslant r} |y|^{\alpha-p}\,\mu\,(dy) = \int_{|y|\geqslant 1} - \int_{B_r\setminus\bar{B}_1}$$

and

$$\int_{B_r\setminus\bar{B}_1} |y|^{\alpha-p}\,\mu\,(dy) \leqslant \mu\,(B_r\setminus\bar{B}_1)$$

which is finite. Hence, if (3.4) does not hold then, for all $r > 1$,

$$\int_{|y|\geqslant r} |y|^{\alpha-p}\,\mu\,(dy) = +\infty$$

Thus, putting $r = |x|+1$, we have $U_\alpha^\mu(x) = +\infty$ in $\mathbb{R}^p$.

Conversely, suppose that (3.4) holds. Now, for $y \neq 0$,

$$\int_{|x|<r} |x-y|^{\alpha-p}\,dx = |y|^{\alpha-p} \int_{|x|<r} \left(\frac{|x-y|}{|y|}\right)^{\alpha-p} dx$$

and

$$\frac{|x-y|}{|y|} \geqslant \frac{|y|-|x|}{|y|} \geqslant \frac{|y|-r}{|y|} \text{ when } |x| < r.$$

The last expression is $\geqslant \frac{1}{2}$ when $|y| > 2r$. Thus

$$\int_{|x| \leqslant r} |x-y|^{\alpha-p}\, dx < (\tfrac{1}{2})^{\alpha-p} |y|^{\alpha-p}\, mB_r(0) \text{ when } |y| > 2r.$$

Suppose now that μ_1 is the restriction of μ to $B_{2r}(0)$ and μ_2 the restriction of μ to $\mathbb{R}^p \setminus B_{2r}(0)$. Then μ_1 is of compact support and

$$U_\alpha^\mu(x) = U_\alpha^{\mu_1}(x) + U_\alpha^{\mu_2}(x).$$

We have shown that $U_\alpha^{\mu_1}(x)$ is finite a.e. so it remains only to show the same to be true of $U_\alpha^{\mu_2}(x)$. Since (3.4) holds

$$\int |y|^{\alpha-p}\, \mu_2\,(dy) < +\infty,$$

and, for every $r > 0$,

$$\int_{|x| \leqslant r} U_\alpha^{\mu_2}(x)\, dx = K \int_{|y| \geqslant 2r} \int_{|x| \leqslant r} |x-y|^{\alpha-p}\, dx\, \mu_2\,(dy)$$

$$\leqslant (\tfrac{1}{2})^{\alpha-p}\, mB_r(0) \int |y|^{\alpha-p}\, \mu_2\,(dy) < +\infty.$$

Hence $U_\alpha^{\mu_2}(x)$ is integrable in $B_r(0)$, so finite a.e. there and hence finite a.e. in $\mathbb{R}^p$.

In what follows we shall assume that any measure considered satisfies condition (3.4). Under this restriction we shall then be able to assume that any α-potential, U_α^γ, of a charge γ, is finite a.e. in $\mathbb{R}^p$.

We turn now to a uniqueness property of α-potentials. This is

THEOREM 3.5. *If $U_\alpha^\gamma(x) = 0$ a.e. in $\mathbb{R}^p$ then γ must be the zero charge.*

We must show that

$$\gamma(\phi) = \int \phi\,(x)\, \gamma(dx) = 0, \quad \phi \in \mathscr{C}_0(\mathbb{R}^p).$$

By Theorem 1.30, and since γ is the difference of two measures, it is enough to show that, for every $\psi \in \mathscr{D}(\mathbb{R}^p)$,

$$\gamma(\psi) = 0.$$

Now, given such a test function ψ, the function $\chi(x) = (k_{-\alpha} * \psi)\,(x)$ is well defined. Furthermore,

$$\chi(x) = 0(|x|^{-\alpha-p}) \quad \text{as } |x| \to +\infty,$$

since $T_1^{(m)}(x)$ is a constant multiple of

$$\int_{B_1(x)} |x-y|^{-\alpha-p}\, (\psi(y) - P_m\,(\psi;\, x, |x-y|)\, dy,$$

which is clearly of this order. Similar considerations apply to $T_2(x)$, bearing in mind that ψ is of compact support. Finally $T_3^{(m)}(x)$ vanishes for large $|x|$. Also $\psi(x) = (k_\alpha * \chi)\,(x)$.

Next,

$$\int_{\mathbb{R}^p} U_\alpha^\gamma(x)\, \chi(x)\, dx = 0$$

and, since $|x-y|^{\alpha-p}|\psi(x)|$ is integrable with respect to each of the product measures $\gamma^+ \times m$ and $\gamma^- \times m$ in $\mathbb{R}^p \times \mathbb{R}^p$ we may, by Fubini's Theorem, invert the order of integration. Thus

$$0 = \int_{\mathbb{R}^p} \int |x-y|^{\alpha-p} \chi(x)\, \gamma(dy)\, dx$$

$$= \int U_\alpha^\chi(y)\, \gamma(dy) = \int \psi(y)\, \gamma(dy).$$

Hence $\gamma(\psi) = 0$ for all test functions ψ and the theorem follows.

§ 3.4. Charges of Finite Energy

The **α-energy** of a charge γ is defined to be

$$\int\int k_\alpha(x-y)\, \gamma\,(dx)\, \gamma\,(dy),$$

whenever this double integral exists. When γ is a measure, its α-energy, finite or infinite, is defined; but when γ is a charge, the integral may or may not exist. When it does and is finite, the charge γ is said to be of **finite α-energy.**

Suppose now that γ and λ are charges such that the integral

$$\int\int k_\alpha(x-y)\, \gamma\,(dx)\, \lambda\,(dy)$$

exists and is finite. The integral is then said to be the **mutual energy** of γ and λ and denoted by (γ, λ). Clearly

$$(\gamma, \lambda) = \int U_\alpha^\gamma(y)\, \lambda\,(dy),$$

and by decomposing the charges γ and λ into differences of measures and applying Fubini's Theorem to each of the four resulting integrals, we find that

$$(\gamma, \lambda) = \int U_\alpha^\lambda(x)\, \gamma\,(dx).$$

It is then clear that the mutual energy (γ, λ) satisfies the relations

$$(\gamma, \lambda) = (\lambda, \gamma); \quad (\gamma_1+\gamma_2, \lambda) = (\gamma_1, \lambda)+(\gamma_2, \lambda)$$

$$(a\gamma, \lambda) = a(\gamma, \lambda), \quad a \in \mathbb{R}.$$

LEMMA 3.6. *The energy of any charge γ of finite energy is non-negative and $(\gamma, \gamma) = 0$ if, and only if, γ is the zero measure.*

Let γ and λ be two charges of finite energy. Now

$$(\gamma, \lambda) = \int\int k_\alpha(x-y)\, \gamma\,(dx)\, \lambda\,(dy)$$

and, applying the Riesz Composition Formula with α and β each replaced by $\frac{1}{2}\alpha$ we have

$$(\gamma, \lambda) = \int\int\int_{\mathbb{R}^p} k_{\frac{1}{2}\alpha}(x-z)\, k_{\frac{1}{2}\alpha}(z-y)\, dz\, \gamma\,(dx)\, \lambda\,(dy).$$

Inverting the order of integration, we have

$$(\gamma, \lambda) = \int_{\mathbb{R}^p} U^{\gamma}_{\frac{1}{2}\alpha}(z)\, U^{\lambda}_{\frac{1}{2}\alpha}(z)\, dz.$$

Consequently,

$$(\gamma, \gamma) = \int_{\mathbb{R}^p} (U^{\gamma}_{\frac{1}{2}\alpha}(z))^2\, dz$$

so that $(\gamma, \gamma) \geqslant 0$.

Further, if $(\gamma, \gamma) = 0$ then $U^{\gamma}_{\frac{1}{2}\alpha}(z) = 0$ a.e. in $\mathbb{R}^p$ whence, by Theorem 3.5, γ is the zero measure.

Thus (γ, λ) is an inner product in the Hilbert space sense. As is usual, we denote (γ, γ) by $\| \gamma \|^2$.

Denote by $\mathscr{E}_\alpha$ the set of all charges of finite α-energy. Then $\mathscr{E}_\alpha$ is a linear space over $\mathbb{R}$ and has an inner product defined on it. **It is therefore a pre-Hilbert space.**

In $\mathscr{E}_\alpha$ we may define convergence in two ways:

We shall say that a sequence $\{\gamma_n\}$, $\gamma_n \in \mathscr{E}_\alpha$ **converges strongly** to $\gamma \in \mathscr{E}_\alpha$ if $\lim\limits_{n\to\infty} \| \gamma_n - \gamma \| = 0$ and we will write $\gamma_u \twoheadrightarrow \gamma$ in this case.

We shall say that $\{\gamma_n\}$ **converges weakly** to $\gamma \in \mathscr{E}$, if

$$\lim_{n\to\infty} (\gamma_n, \lambda) = (\gamma, \lambda)$$

for all $\lambda \in \mathscr{E}_\alpha$. In this case we will write $\gamma_n \to \gamma$.

Since, in a pre-Hilbert space the Cauchy–Schwarz inequality†

$$| (\gamma, \lambda) | \leqslant \| \gamma \| \, \| \lambda \|$$

holds we have

$$| (\gamma_n, \lambda) - (\gamma, \lambda) | \leqslant \| \gamma_n - \gamma \| \, \| \lambda \|$$

and so,

$$\text{if } \gamma_n \twoheadrightarrow \gamma, \text{ then } \gamma_n \to \gamma,$$

so that **strong convergence implies weak convergence.**

Finally, we shall need the concept of vague convergence for charges. Given a sequence $\{\gamma_n\}$ of charges and a charge γ such that

$$\lim_{n\to\infty} \gamma_n(f) = \gamma(f), \quad f \in \mathscr{C}_0(\mathbb{R}^p)$$

we say that γ_n converges vaguely to γ and write $\gamma_n \rightsquigarrow \gamma$.

Then **weak convergence implies vague convergence.**

We have

$$(\gamma_n - \gamma, \lambda) = \int U^{\gamma_n}_{\alpha}(x)\, \lambda\, (dx) - \int U^{\gamma}_{\alpha}(x)\, \lambda\, (dx)$$

$$= \int U^{\lambda}_{\alpha}(x)\, \gamma_n\, (dx) - \int U^{\lambda}_{\alpha}(x)\, \gamma\, (dx)$$

† Consider $\| \gamma + a\lambda \|^2 = \| \gamma \|^2 + 2a(\gamma, \lambda) + a^2 \| \lambda \|^2$. The quadratic in a is non-negative for all $a \in \mathbb{R}$, and so the inequality follows.

and so, if $\gamma_n \to \gamma$, then

$$\lim_{n\to\infty} \int U_\alpha^\lambda(x)\, \gamma_n(dx) = \int U_\alpha^\lambda(x)\, \gamma(dx) \quad \text{for } \lambda \in \mathscr{E}_\alpha.$$

Let $\phi \in \mathscr{D}(\mathbb{R}^p)$. Then, by Lemma 3.2, we can find an infinitely differentiable function ψ with $\psi(x) = 0(|x|^{-\alpha-p})$ for large $|x|$ such that $\phi = U_\alpha^\psi$. Also, the charge defined by

$$\psi(f) = \int \psi(x) f(x)\, dx, \quad f \in \mathscr{C}_0(\mathbb{R}^p)$$

where, as is usual, we denote the charge associated with the function by ψ also, belongs to $\mathscr{E}_\alpha$ since

$$\int\int |x-y|^{\alpha-p} |\psi(y)|\, |\psi(x)|\, dy\, dx < +\infty. \tag{3.5}$$

To see that (3.5) holds, we proceed as follows: For $x \neq 0$,

$$\int |x-y|^{\alpha-p} |\psi(y)|\, dy = I_1 + I_2 + I_3$$

where these denote the integrals over the ranges $|y| < \frac{1}{2}|x|$, $\frac{1}{2}|x| \leqslant |y| \leqslant \frac{3}{2}|x|$ and $|y| > \frac{3}{2}|x|$ respectively. Now

$$I_1 \leqslant (\tfrac{1}{2}|x|)^{\alpha-p} \int_{|y| \leqslant \frac{1}{2}|x|} |\psi(y)|\, dy$$

and this last integral is

$$\leqslant \int_{|y|\leqslant 1} |\psi(y)|\, dy + C \int_{|y|\geqslant 1} |y|^{-\alpha-p}\, dy < +\infty.$$

Thus $I_1 = 0(|x|^{\alpha-p})$.

Next

$$I_2 \leqslant C|x|^{-p} \int_{\frac{1}{2}\leqslant |z| \leqslant \frac{3}{2}} |w-z|^{\alpha-p} |z|^{-\alpha-p}\, dz, \quad w = x/|x|$$

and the integral is

$$\leqslant (\tfrac{1}{2})^{-\alpha-p} \int_{-\frac{1}{2}\leqslant |u| \leqslant \frac{3}{2}} |u|^{\alpha-p}\, du$$

so that $I_2 = 0(|x|^{-p})$.

Finally,

$$I_3 \leqslant C|x|^{-p} \int_{|z|\geqslant\frac{3}{2}} |w-z|^{\alpha-p} |z|^{-\alpha-p}\, dz$$

$$< (\tfrac{1}{2})^{\alpha-p} |x|^{-p} \int_{\frac{3}{2}}^{\infty} \rho^{-\alpha-1}\, d\rho < +\infty;$$

whence $I_3 = 0(|x|^{-p})$.

Consequently, the double integral in (3.5) is

$$\leqslant \int |x|^{\alpha-p} |\psi(x)|\, dx \leqslant \int_{|x|\leqslant 1} |x|^{\alpha-p} |\psi(x)|\, dx + C\int |x|^{-2p}\, dx < +\infty.$$

Thus

$$\lim_{n\to\infty} \int U_\alpha^\psi(x)\, \gamma_n\,(dx) = \int U_\alpha^\psi(x)\, \gamma\,(dx).$$

But, since $U_\alpha^\psi = \phi$ we then have

$$\lim_{n\to\infty} \gamma_n(\phi) = \gamma(\phi)$$

and this for every $\phi \in \mathscr{D}(G)$. Hence $\gamma_n \rightsquigarrow \gamma$, by Theorem 1.30.

We turn next to the proof of two density theorems for $\mathscr{E}_\alpha$. The crux of the proof for the two results, Theorems 3.10 and 3.11 is, in fact, contained in Lemma 3.9, concerning the behaviour of the α-th integral of the function τ_δ. Lemmas 3.7 and 3.8 are merely preliminaries to Lemma 3.9.

LEMMA 3.7. *Suppose that* $0 < a < 1$ *and* $b > -1$. *Then*

$$\int_0^\infty \left|\frac{k}{t-k}\right|^a t^b\, e^{-t}\, dt$$

is bounded as k ranges over $\mathbb{R}$.

Suppose first that $k > 0$ and set $t = ku$. The integral becomes

$$k^{b+1}\left(\int_0^{\frac{1}{2}} + \int_{\frac{1}{2}}^{\frac{3}{2}} + \int_{\frac{3}{2}}^{\infty}\right) |\,u-1\,|^{-a}\, u^b\, e^{-ku}\, du.$$

The first and third integrals are less than $2^a\, \Gamma(b+1)k^{-b-1}$ and the second is less than

$$e^{-\frac{1}{2}k} \int_{\frac{1}{2}}^{\frac{3}{2}} |\,u-1\,|^{-a}\, u^b\, du,$$

which gives the result.

When $k < 0$ the substitution $t = |\,k\,|\, u$ gives the integral

$$|\,k\,|^{b+1} \int_0^\infty |\,u+1\,|^{-a}\, u^b\, e^{-|k|u}\, du < |\,k\,|^{b+1} \int_0^\infty |\,u-1\,|^{-a}\, u^b\, e^{-|k|u}\, du,$$

so, by what has just been proved, the result again follows.

LEMMA 3.8. *Suppose that* $0 < a < 1$ *and b is real. Then*

$$\int_{2k}^\infty \left|\frac{\sqrt{(kt)}}{t-k}\right|^a t^b\, e^{-t}\, dt \tag{3.6}$$

is bounded as k ranges over $(0, \infty)$.

Let $t = k(u+2)$. Then the integral becomes

$$k^{b+1}\, e^{-2k} \int_0^\infty \left(\frac{u+2}{u+1}\right)^a (u+2)^{b-\frac{1}{2}a}\, e^{-ku}\, du.$$

This last integral is less than

$$2^a \int_0^\infty (u+2)^{b-\frac{1}{2}a}\, e^{-ku}\, du.$$

When $b-\frac{1}{2}a > 0$, this last is less than a multiple of

$$\int_0^\infty u^{b-\frac{1}{2}a}\, e^{-ku}\, du + \int_0^\infty e^{-ku}\, du = \Gamma(b-\tfrac{1}{2}a+1)\, k^{\frac{1}{2}a-b-1} + k^{-1}$$

and so (3.6) is of order $e^{-2k}\, k^{\frac{1}{2}a} + k^b\, e^{-2k}$, which is bounded.

When $-1 < b-\frac{1}{2}a \leqslant 0$, the integral is less than a multiple of

$$\int_0^\infty u^{b-\frac{1}{2}a}\, e^{-ku}\, du = \Gamma(b-\tfrac{1}{2}a+1)\, k^{\frac{1}{2}a-b-1}$$

and so, once again, (3.6) is bounded.

When $b-\frac{1}{2}a \leqslant -1$, the integral is less than a multiple of

$$\int_0^2 e^{-ku}\, du + \int_2^\infty u^{b-\frac{1}{2}a}\, e^{-ku}\, du.$$

The first integral is bounded for all k and, for $b-\frac{1}{2}a < -1$, so is the second. When $b-\frac{1}{2}a = -1$, the second integral becomes

$$\int_2^\infty u^{-1}\, e^{-ku}\, du = \int_{2k}^\infty t^{-1}\, e^{-t}\, dt,$$

which is bounded for large k and is 0 (log k) for small k. It follows that (3.6) is bounded.

LEMMA 3.9. *Let* $0 < \alpha < p$, $0 \leqslant \delta \leqslant 1$. *Then*

$$U_\alpha^{\tau_\delta}(x) \leqslant Ck_\alpha(x)$$

where C *is a constant independent of* δ.

Because of Lemma 1.27 this amounts to proving that

$$\int_{|y|<\delta} |\, x-y\, |^{\alpha-p}\, e^{-1/(\delta^2-|y|^2)}\, dy < C\delta^{2p}\, e^{-1/\delta^2}\, |\, x\, |^{\alpha-p}. \tag{3.7}$$

It is convenient to treat the cases $p = 1$ and $p \geqslant 2$ separately, and we turn first to the case $p = 1$. We must then show that, for $0 < \alpha < 1$,

$$\int_{-\delta}^{\delta} |\, \xi-\eta\, |^{\alpha-1}\, e^{-1/(\delta^2-\eta^2)}\, d\eta < C\delta^2\, e^{-1/\delta^2}\, |\, \xi\, |^{\alpha-1}.$$

We may suppose that $\xi > 0$, and then the integral on the left is less than twice that over the range $[0, \delta]$. In this last set $\eta = \delta^2\sqrt{t}/\sqrt{(1+\delta^2 t)}$. The resulting integral is less than

$$\tfrac{1}{2}\delta^2 e^{-1/\delta^2} \xi^{\alpha-1} \int_0^\infty \left| 1-\frac{\delta^2\sqrt{t}}{\xi\sqrt{(1+\delta^2 t)}} \right|^{\alpha-1} t^{-\frac{1}{2}} e^{-t}\, dt. \tag{3.8}$$

When $\xi \geqslant \delta$

$$1-\frac{\delta^2\sqrt{t}}{\xi\sqrt{(1+\delta^2 t)}} \geqslant 1-\frac{\delta\sqrt{t}}{\sqrt{(1+\delta^2 t)}}$$

$$= 1/(1+\delta^2 t+\delta\sqrt{t}+\delta^2 t^2) \geqslant \tfrac{1}{2}(1+t)^{-1}$$

and so (3.8) is less than

$$2^{-\alpha}\, \delta^2\, e^{-1/\delta^2}\, \xi^{\alpha-1} \int_0^\infty (1+t)^{\alpha-1}\, t^{-\frac{1}{2}}\, e^{-t}\, dt$$

and we have the result for $p = 1$ and $\xi \geqslant \delta$.

When $0 < \xi < \delta$ let $t_0 = \xi^2/\delta^2(\delta^2-\xi^2)$. Then

$$\left| 1-\frac{\delta^2\sqrt{t}}{\xi\sqrt{(1+\delta^2 t)}} \right|^{-1}$$

$$= (1+\delta^2 t)\,\{t_0+(tt_0)^{\frac{1}{2}}\,[(1+\delta^2 t_0)/(1+\delta^2 t)]^{\frac{1}{2}}\}/|\, t-t_0\,|$$

Now

$$(tt_0)^{\frac{1}{2}}\,[(1+\delta^2 t_0)/(1+\delta^2 t)]^{\frac{1}{2}} = t_0\left(\frac{\delta^2+1/t_0}{\delta^2+1/t}\right)^{\frac{1}{2}}$$

and this is not greater than t_0 when $t \leqslant t_0$ and is not greater than $t_0\sqrt{(t/t_0)}$ when $t \geqslant t_0$. Consequently

$$|\,1-\eta/\xi\,|^{\alpha-1} < \begin{cases} (1+t^{1-\alpha})\left|\dfrac{(1+\sqrt{2})\,t_0}{t-t_0}\right|^{1-\alpha} & \text{when } t \leqslant 2t_0 \\[2ex] (1+t^{1-\alpha})\left\{\left|\dfrac{t_0}{t-t_0}\right|^{1-\alpha}+\left|\dfrac{\sqrt{(tt_0)}}{t-t_0}\right|^{1-\alpha}\right\} & \text{when } t > 2t_0 \end{cases}$$

and, in consequence, the integral in (3.8) is less than a multiple of the sum of the four integrals

$$\int_0^\infty \left|\frac{t_0}{t-t_0}\right|^{1-\alpha} t^b\, e^{-t}\, dt,\quad \int_{2t_0}^\infty \left|\frac{\sqrt{(tt_0)}}{t-t_0}\right|^{1-\alpha} t^b\, e^{-t}\, dt,\; b = -\tfrac{1}{2}, \tfrac{1}{2}-\alpha.$$

By Lemmas 3.7 and 3.8 each of these has a bound independent of t_0. We have thus dealt with the case $p = 1$.

We turn now to the case $p \geqslant 2$. Choose coordinates so that $x = (\xi, 0, \ldots, 0)$ with $\xi > 0$. Then the integral in (3.7) becomes

$$\int_{y_1^2+\ldots+y_p^2 \leqslant \delta^2} [(\xi-y_1)^2+y_2^2+\ldots+y_p^2]^{\frac{1}{2}(\alpha-p)} \exp(-[\delta^2-y_1^2-\ldots-y_p^2]^{-1})\, dy_1 \ldots dy_p.$$

Transform $y_2, \ldots, y_p$ to $(p-1)$-dimensional spherical coordinates to obtain

$$\frac{\pi^{\frac{1}{2}(p-1)}}{\Gamma(\frac{1}{2}(p-1))} \int_0^\delta \int_0^{\delta^2-\eta^2} [(\xi-\eta)^2+\rho^2]^{\frac{1}{2}(\alpha-p)} \exp(-[\delta^2-\eta^2-\rho^2]^{-1})\, \rho^{p-2}\, d\rho\, d\eta. \tag{3.9}$$

Setting $k = \sqrt{(\delta^2-\eta^2)}$, $q = \xi-\eta$, and making the substitution $\rho = k^2\sqrt{s}/\sqrt{(1+k^2 s)}$, the inner integral in (3.9) becomes

$$\tfrac{1}{2}k^{2p-2}\, e^{-1/k^2} \int_0^\infty [q^2+k^2(q^2+k^2)s]^{\frac{1}{2}(\alpha-p)} (1+k^2 s)^{-\frac{1}{2}(\alpha+1)}\, s^{\frac{1}{2}(p-3)}\, e^{-s}\, ds$$

and hence (3.9) is less than a multiple of

$$\int_0^\infty s^{\frac{1}{2}(p-3)}\, e^{-s}\, ds \int_0^\delta [(\xi-\eta)^2+(\delta^2-\eta^2)(\xi^2-2\xi\eta+\delta^2)s]^{\frac{1}{2}(\alpha-p)} (\delta^2-\eta^2)^{p-1}\, e^{-1/(\delta^2-\eta^2)}\, d\eta.$$

In the inner integral we make the substitution $\eta = \delta^2\sqrt{t}/\sqrt{(1+\delta^2 t)}$ and the integral becomes

$$\tfrac{1}{2}\delta^{2p}\, e^{-1/\delta^2}\, \xi^{\alpha-\frac{1}{2}p} \int_0^\infty s^{\frac{1}{2}(p-3)}\, e^{-s}\, ds \int_0^\infty (1+\delta^2 t)^{-p-\frac{1}{2}} (F(t))^{\frac{1}{2}(\alpha-p)}\, t^{-\frac{1}{2}}\, e^{-t}\, dt \tag{3.10}$$

where

$$F(t) = \left(1-\frac{\delta}{\xi}\frac{\delta\sqrt{t}}{\sqrt{(1+\delta^2 t)}}\right)^2 + \frac{\delta^2 s}{1+\delta^2 t}\left(1-2\frac{\delta}{\xi}\frac{\delta\sqrt{t}}{\sqrt{(1+\delta^2 t)}}+\left(\frac{\delta}{\xi}\right)^2\right)$$

Suppose first that $\xi \geqslant \delta$. Then

$$1-\frac{\delta}{\xi}\frac{\delta\sqrt{t}}{\sqrt{(1+\delta^2 t)}} \geqslant \tfrac{1}{2}(1+t)^{-1}$$

and so (3.10) is less than a multiple of

$$\delta^{2p}\, e^{-1/\delta^2}\, \xi^{\alpha-p} \int_0^\infty \int_0^\infty s^{\frac{1}{2}(p-3)}\, e^{-s}(1+t)^{\alpha-p}\, t^{-\frac{1}{2}}\, e^{-t}\, dt\, ds,$$

which thus gives the result for $\xi \geqslant \delta$ when $p \geqslant 2$.

Otherwise, suppose that $0 < \xi < \delta$. By the equality

$$(1-u\lambda)^2+v(1-2u\lambda+u^2) = \frac{[(\lambda^2+v)u-\lambda(1+v)]^2}{\lambda^2+v}+\frac{v(1+v)\,(1-\lambda^2)}{\lambda^2+v}$$

with $u = \delta/\xi$, $\lambda = \delta\sqrt{t}/\sqrt{(1+\delta^2 t)}$ and $v = \delta^2 s/(1+\delta^2 t)$ we find that

$$F(t) = A+B,$$

where

$$A = [(1+\delta^2 t)\,(s+t)]^{-1}\left[(s+t)\,\frac{\delta^2}{\xi}-\frac{\sqrt{t}}{\sqrt{(1+\delta^2 t)}}\,(1+\delta^2(s+t))\right]^2$$

$$B = \frac{s}{s+t}\;\frac{1+\delta^2(s+t)}{(1+\delta^2 t)^2}.$$

Let $s_t = \xi\sqrt{t}\,\delta^{-2}\,(\sqrt{(1+\delta^2 t)}-\xi\sqrt{t})^{-1}-t$. Then

$$\xi^{-1} = \frac{\sqrt{t}}{\sqrt{(1+\delta^2 t)}}\;\frac{1+\delta^2(s_t+t)}{\delta^2(s_t+t)},$$

and hence

$$A = t(1+\delta^2 t)^{-2}\,(s+t)^{-1}\,[(s-s_t)/(s_t+t)]^2.$$

Since $A+B \geqslant A^{1/p}\,B^{1-1/p}$ we have

$$(A+B)^{\frac{1}{2}(\alpha-p)} \leqslant A^{(\alpha-p)/2p}\;B^{(\alpha-p)(p-1)/2p}$$

and so the inner integral in (3.10) is less than

$$\int_0^\infty (1+\delta^2 t)^{-\alpha-\frac{1}{2}}\,(s+t)^{\frac{1}{2}(p-\alpha)}\,s^{(\alpha-p)(p-1)/2p}\left|\frac{s-s_t}{t+s_t}\right|^{(\alpha-p)/p}$$

$$(1+\delta^2(s+t))^{(\alpha-p)(p-1)/2p}\;t^{\alpha/2p-1}\;e^{-t}\,dt.$$

Since

$$(s+t)^{\frac{1}{2}(p-\alpha)} \leqslant 2^{\frac{1}{2}(p-\alpha)}\,(s^{\frac{1}{2}(p-\alpha)}+t^{\frac{1}{2}(p-\alpha)})$$

we have that (3.10) itself is less than a multiple of

$$\delta^{2p}\,e^{-1/\delta^2}\,\xi^{\alpha-p}\,[\int_0^\infty (t+s_t)^{(p-\alpha)/p}\,t^{\alpha/2p-1}\,e^{-t}\,dt$$

$$\int_0^\infty s^{\frac{p-1}{2p}\alpha-1}\,(s^{\frac{1}{2}(p-\alpha)}+t^{\frac{1}{2}(p-\alpha)})\,|\,s-s_t\,|^{(\alpha-p)/p}\,e^{-s}\,ds].$$

By application of Lemma 3.7 to the inner integral this last is less than a multiple of

$$\delta^{2p}\,e^{-1/\delta^2}\,\xi^{\alpha-p}\left[\int_0^\infty\left|\frac{t+s_t}{s_t}\right|^{(p-\alpha)/p} t^{\alpha/2p-1}\,e^{-t}\,dt\right.$$

$$\left.+\int_0^\infty\left|\frac{t+s_t}{s_t}\right|^{(p-\alpha)/p} t^{\alpha\frac{1-p}{2p}+\frac{p-2}{2}}\,e^{-t}\,dt\right]. \qquad (3.11)$$

Now

$$\frac{s_t}{t+s_t} = 1+\delta^2 t-\delta^2\sqrt{(t+\delta^2 t^2)}/\xi.$$

Let $t_0 = \xi^2/\delta^2(\delta^2-\xi^2)$ so that $\xi^{-1} = \sqrt{(1+\delta^2 t_0)}/\delta^2\sqrt{t_0}$ and hence

$$\frac{t+s_t}{s_t} = \left(t_0+\sqrt{(tt_0)}\sqrt{\left(\frac{1+\delta^2 t_0}{1+\delta^2 t}\right)}\right)\Big/(t_0-t)$$

and so

$$\left|\frac{t+s_t}{s_t}\right| \leqslant \begin{cases} (1+\sqrt{2})t_0/|t-t_0| & \text{when } t \leqslant 2t_0 \\ \dfrac{t_0}{|t-t_0|}+\dfrac{\sqrt{(tt_0)}}{|t-t_0|} & \text{when } t > 2t_0. \end{cases}$$

Thus the first integral in the square bracket in (3.11) is less than a multiple of

$$\int_0^\infty \left|\frac{t_0}{t-t_0}\right|^{(p-\alpha)/p} t^{\alpha/2p-1}\, e^{-t}\, dt+\int_{2t_0}^\infty \left|\frac{\sqrt{(tt_0)}}{t-t_0}\right|^{(p-\alpha)/p} t^{\alpha/2p-1}\, e^{-t}\, dt.$$

Using Lemma 3.7 for the first of these integrals and Lemma 3.8 for the second, the first integral has a bound independent of t_0. A similar argument shows that this also holds for the second integral in the square bracket.

This completes the proof of Lemma 3.9.

THEOREM 3.10. *The set of infinitely differentiable functions is strongly dense in* $\mathscr{E}_\alpha$.

Let $\gamma \in \mathscr{E}_\alpha$. It is enough to show that, if $\gamma_n = \gamma * \tau_{1/n} = \int \tau_{1/n}(x-y)\,\gamma(dy)$, then γ_n converges strongly to γ. First, for $0 < \beta < p$,

$$\limsup_{n\to\infty} U_\beta^{\tau_{1/n}}(x) \leqslant k_\beta(x). \tag{3.12}$$

For, if $x = 0$, this is trivial. If $x \neq 0$ then, for $n > 1/|x|$,

$$\int |x-y|^{\beta-p}\,\tau_{1/n}(y)\,dy = \int_{|y|\leqslant 1/n} \leqslant [|x|-(1/n)]^{\beta-p} \int \tau_{1/n}(y)\,dy$$
$$= [|x|-(1/n)]^{\beta-p}$$

and, letting $n\to\infty$, (3.12) follows.

Furthermore,

$$\liminf_{n\to\infty} U_\beta^{\tau_{1/n}}(x) \geqslant k_\beta(x). \tag{3.13}$$

For let $f_m(z) = \min [m, k_\beta(z)]$. Then f_m tends increasingly to k_β as n tends to infinity and so

$$U_\beta^{\tau_{1/n}}(x) \geqslant \int f_m(x-y)\, \tau_{1/n}(y)\, dy.$$

Since f_m is continuous, the right hand side tends to $f_m(x)$ as n tends to infinity, by Theorem 1.28(*ii*), and thus

$$\liminf_{n\to\infty} U_\beta^{\tau_{1/n}}(x) \geqslant f_m(x) \quad \text{for every } m.$$

From this (3.13) follows and from (3.12) and (3.13) we have

$$\lim_{n\to\infty} U_\beta^{\tau_{1/n}}(x) = k_\beta(x). \tag{3.14}$$

Let $\gamma = \gamma^+ - \gamma^-$, where these are the positive and negative variations of γ. Then $\gamma_n = \gamma_n^+ - \gamma_n^-$ where $\gamma_n^+ = \gamma^+ * \tau_{1/n}$ and γ_n^- is similarly related to γ^- and, if γ_n^+ and γ_n^- converge strongly to γ^+ and γ^-, then γ_n converges strongly to γ. We may therefore assume that γ is a measure. Now

$$\| \gamma_n - \gamma \|^2 = \int \int k_\alpha(x-y)\, (\gamma_n - \gamma)\, (dx)\, (\gamma_n - \gamma)\, (dy)$$

and, using the Riesz Composition Formula, this last becomes

$$\int \int \int_{\mathbb{R}^p} k_{\frac{1}{2}\alpha}(x-z)\, k_{\frac{1}{2}\alpha}(z-y)\, dz (\gamma_n - \gamma)\, (dx)\, (\gamma_n - \gamma)\, (dy).$$

Changing the order of integration, this may be rewritten as

$$\int (U_{\frac{1}{2}\alpha}^{\gamma_n - \gamma}(z))^2\, dz = \int (U_{\frac{1}{2}\alpha}^{\gamma_n}(z) - U_{\frac{1}{2}\alpha}^{\gamma}(z))^2\, dz. \tag{3.15}$$

Also

$$\begin{aligned} U_{\frac{1}{2}\alpha}^{\gamma_n}(z) &= U_{\frac{1}{2}\alpha}^{\gamma * \tau_{1/n}}(z) = \int k_{\frac{1}{2}\alpha}(z-x)\, (\gamma * \tau_{1/n})\, (dx) \\ &= \int k_{\frac{1}{2}\alpha}(z-x)\, dx \int \tau_{1/n}(x-y)\, \gamma\, (dy) = \int U_{\frac{1}{2}\alpha}^{\tau_{1/n}}(z-y)\, \gamma\, (dy). \end{aligned}$$

Now $\gamma \in \mathscr{E}_\alpha$ and so $(\gamma, \gamma) = \int_{\mathbb{R}^p} (U_{\frac{1}{2}\alpha}^{\gamma}(z))^2\, dz$ is finite. Thus

$$U_{\frac{1}{2}\alpha}^{\gamma}(z) = \int k_{\frac{1}{2}\alpha}(z-y)\gamma(dy)$$

is finite for z a.e. Consequently, $k_{\frac{1}{2}\alpha}(z-y)$ is, as a function of y, γ-integrable for z a.e. Then, by Lemma 3.9 and by (3.14) $U_{\frac{1}{2}\alpha}^{\tau_{1/n}}(z-y)$ converges, as $n \to \infty$, dominatedly to $k_{\frac{1}{2}\alpha}(z-y)$, qua y, for z a.e. Hence, by Theorem 1.16,

$$\lim_{n\to\infty} U_{\frac{1}{2}\alpha}^{\gamma_n}(z) = U_{\frac{1}{2}\alpha}^{\gamma}(z).$$

Furthermore, again by Lemma 3.9, $U_{\frac{1}{2}\alpha}^{\gamma_n}(z) < C\, U_{\frac{1}{2}\alpha}^{\gamma}(z)$, and so

$$(U_{\frac{1}{2}\alpha}^{\gamma_n}(z) - U_{\frac{1}{2}\alpha}^{\gamma}(z))^2 \leqslant 2(U_{\frac{1}{2}\alpha}^{\gamma_n}(z))^2 + 2(U_{\frac{1}{2}\alpha}^{\gamma}(z))^2 \leqslant 2(1+C^2)\, (U_{\frac{1}{2}\alpha}^{\gamma}(z))^2.$$

Consequently, the integrand in (3.15) converges dominatedly to 0. Thus $\| \gamma_n - \gamma \|$ tends to 0 as n tends to infinity, and Theorem 3.10 is proved.

THEOREM 3.11. *The set of charges in $\mathscr{E}_\alpha$ of which the α-potentials are test functions is strongly dense in $\mathscr{E}_\alpha$.*

First, *the set of charges of compact support is strongly dense in $\mathscr{E}_\alpha$.* For let $\gamma \in \mathscr{E}_\alpha$ and let γ_n be the restriction of γ to the ball B_n of centre 0 and radius n. Then

$$\| \gamma_n - \gamma \|^2 = \int \int k_\alpha(x-y)\,(\gamma_n-\gamma)\,(dx)\,(\gamma_n-\gamma)\,(dy)$$
$$= \int_{\mathbb{R}^p \setminus B_n} \int_{\mathbb{R}^p \setminus B_n} k_\alpha(x-y)\,\gamma\,(dx)\,\gamma\,(dy)$$
$$\leqslant \int_{\mathbb{R}^{2p} \setminus B_n^*} k_\alpha(x-y)\,\gamma\,(dx)\,\gamma\,(dy),$$

where B_n^* is the ball of radius n in $\mathbb{R}^{2p}$ centred at 0. Since the integral over $\mathbb{R}^{2p}$ is finite, this last converges to zero as n tends to infinity.

When γ is of compact support, $\gamma * \tau_{1/n}$ is a test function; and so the set of test functions is dense in the set of charges in $\mathscr{E}_\alpha$ of compact support. Hence, to prove the theorem it will be enough to show that a test function charge can be approximated by charges of which the potentials are test functions.

Let, then, γ be a test function. Since it is of compact support, we may choose $R > 0$ so that supp $\gamma \subset B_R$. Let χ_ρ be the indicator function for B_ρ. Then $\chi_{3R} * \tau_R$ lies between 0 and 1, takes the value 1 in B_{2R} and the value 0 outside B_{4R} and is a test function. Let

$$f(x) = (U_\alpha^\gamma(x))\,(\chi_{3R} * \tau_R)\,(x).$$

Since U_α^γ is infinitely differentiable, f is a test function. We can therefore, by Lemma 3.2, find an infinitely differentiable function ψ such that $f = U_\alpha^\psi$. Also $\psi(x) = 0(|x|^{-\alpha-p})$ and so, as we have seen in the course of the proof of Theorem 3.5, $\psi \in \mathscr{E}_\alpha$.

Consider

$$F(x) = f(x) - U_\alpha^\gamma(x) = U_\alpha^\psi(x) - U_\alpha^\gamma(x).$$

Since $f(x) = U_\alpha^\gamma(x)$ in B_{2R} we have $F(x) = 0$ in B_{2R}. Also supp $\gamma \subset B_R$ and so

$$\| \psi - \gamma \|^2 = \int F(x)\,(\psi(x) - \gamma(x))\,dx = \int_{\mathbb{R}^p \setminus B_{2R}} F(x)\,\psi(x)\,dx.$$

Now $U_\alpha^\gamma(x) = 0(|x|^{\alpha-p})$ in $\mathbb{R}^p \setminus B_{2R}$, and so this holds for $F(x)$. Also $\psi(x) = 0(|x|^{-\alpha-p})$, and so

$$\left| \int_{\mathbb{R}^p \setminus B_{2R}} F(x)\,\psi(x)\,dx \right| < C \int_{2R}^{\infty} \rho^{-2p}\,\rho^{p-1}\,d\rho = C(2R)^{-p}$$

and hence $\| \psi - \gamma \|$ tends to zero as R tends to infinity. Since ψ is a charge of which the α-potential is a test function, the theorem now follows.

A sequence $\{\gamma_n\}$, $\gamma_n \in \mathscr{E}_\alpha$, such that, given $\varepsilon > 0$, there is an N such that, for $m, n > N$,

$$\| \gamma_m - \gamma_n \| < \varepsilon$$

is said to be a **Cauchy sequence** in $\mathscr{E}_\alpha$.

We have seen that $\mathscr{E}_\alpha$ is a pre-Hilbert space and the question now arises as to whether it is a Hilbert space. It would be, if it were complete, that is, if any Cauchy sequence in $\mathscr{E}_\alpha$ converged strongly to an element of $\mathscr{E}_\alpha$. Henri Cartan has produced an example to show that $\mathscr{E}_2$ is not complete. This example also serves to show that $\mathscr{E}_\alpha$ for $\alpha > 1$ is not complete as we shall demonstrate presently. It seems highly probable that $\mathscr{E}_\alpha$ for $\alpha \leqslant 1$ is not complete but Cartan's example fails in this case and the question still seems to be open.

We shall show first that $U_\alpha^{\sigma_r{}^0}(x)$ is, regarded as a function of r, continuous uniformly for all $x \in \mathbb{R}^p$ when $\frac{1}{2} \leqslant r \leqslant \frac{3}{2}$.

Choose coordinates so that $x = (\rho, 0, \ldots, 0)$ where $\rho = |x|$. Then $U_\alpha^{\sigma_r{}^0}(x)$ is a constant multiple of

$$\int_0^\pi (\rho^2 - 2\rho r \cos\theta + r^2)^{\frac{1}{2}(\alpha-p)} \sin^{p-2}\theta \, d\theta.$$

For $\frac{1}{2} \leqslant r \leqslant \frac{3}{2}$,

$$\rho^2 - 2\rho r \cos\theta + r^2 = (\rho - r)^2 + 4\rho r \sin^2 \tfrac{1}{2}\theta \geqslant \tfrac{1}{4}\sin^2 \tfrac{1}{2}\theta,$$

and so the integrand above is dominated by $2^{p-\alpha} \sin^{\alpha-p} \frac{1}{2}\theta \sin^{p-2}\theta$ which is integrable when $\alpha > 1$. It follows, by Theorem 1.23, that $U_\alpha^{\sigma_r{}^0}(x)$ is continuous in the set $\{(\rho, r) \mid \rho \in \mathbb{R}, \frac{1}{2} \leqslant r \leqslant \frac{3}{2}\}$. If we adjoin the point ω to this subset of $\mathbb{R}^2$, the resulting set is compact. Since $(\rho^2 - 2\rho r \cos\theta + r^2)^{\frac{1}{2}(\alpha-p)} \to 0$ as $(\rho, r) \to \omega$ in this set, it is continuous and so uniformly continuous in this compact set. Consequently $U_\alpha^{\sigma_r{}^0}(x)$ is continuous in r uniformly for $x \in \mathbb{R}^p$.

Define the sequence $\{\beta_n\}$ of charges by

$$\beta_n = n\sigma_1 - \sum_{k=1}^n \sigma_{r_k}$$

where σ_r denotes σ_r^0 and $\{r_n\}$ is a strictly increasing sequence converging to 1. Now $\beta_n - \beta_{n-1} = \sigma_1 - \sigma_{r_n}$, and we may choose the sequence $\{r_n\}$ so that

$$| U_\alpha^{\beta_n - \beta_{n-1}}(x) | = | U_\alpha^{\sigma_1}(x) - U_\alpha^{\sigma_{r_n}}(x) | < 2^{-n}.$$

Consequently

$$\| \beta_n - \beta_{n-1} \|^2 = \int U_\alpha^{\beta_n - \beta_{n-1}}(x)\, (\sigma_1 - \sigma_{r_n})\, (dx) < 2^{1-n}$$

and so, for $m > n$,

$$\| \beta_m - \beta_n \| \leqslant \sum_{k=n+1} \| \beta_k - \beta_{k-1} \| < \sum_{n-1}^{\infty} (\sqrt{2})^{1-k}$$

and so $\{\beta_n\}$ is a Cauchy sequence in $\mathscr{E}_\alpha$.

Suppose, if possible, that there is a charge β such that $\beta_n \twoheadrightarrow \beta$. Then, since strong convergence implies vaguec onvergence, we may, for $0 < \rho < 1$, choose a continuous function f taking the value 1 on $B_\rho(0)$, the value 0 outside $B_{\frac{1}{2}(1+\rho)}(0)$ and such that $0 \leqslant f(x) \leqslant 1$ everywhere, for which

$$\lim_{n\to\infty} \int f(x)\, \beta_n\,(dx) = \int f(x)\, \beta(dx).$$

Then

$$\beta(B_\rho) + \int_{\rho<|x|<\frac{1}{2}(1+\rho)} f(x)\, \beta(dx) < -\sum_{r_k \leqslant \rho} 1$$

and so

$$-\beta^-(B_\rho) - \int_{\rho<|x|<\frac{1}{2}(\rho+1)} f(x)\, \beta^-(dx) < -\sum_{r_k \leqslant \rho} 1$$

and

$$\beta^-(B_{\frac{1}{2}(1+\rho)}) \geqslant \beta^-(B_\rho) + \int_{\rho<|x|<\frac{1}{2}(1+\rho)} f(x)\, \beta^-(dx) > \sum_{r_k \leqslant \rho} 1.$$

If we now let $\rho \to 1$, we find that $\beta^-(B_1) = \infty$; which is impossible if β is a charge. Thus the sequence $\{\beta_n\}$, although Cauchy, does not have a strong limit. So, **for $\alpha > 1$, $\mathscr{E}_\alpha$ is not complete.**

But, on the other hand, the subset $\mathscr{E}_\alpha^+$ consisting of all the **measures** in $\mathscr{E}_\alpha$ is, in fact, complete. Of course, it is not a Hilbert space, but the fact that it is complete has far-reaching consequences, and we now set out to give a proof of this important result. We do this by way of a number of lemmas.

LEMMA 3.12. *Suppose given $\{\mu_n\}$, $\mu_n \in \mathscr{E}_\alpha^+$ and that $\mu_n \rightsquigarrow \mu$. If, also, $\{\| \mu_n \|\}$ is bounded, then $\mu \in \mathscr{E}_\alpha^+$ and $\mu_n \to \mu$.*

First, $|x-y|^{\alpha-p}$ is l.s.c. in $\mathbb{R}^{2p}$ and, by Lemma 1.32, $\mu_n \times \mu_n \rightsquigarrow \mu \times \mu$. So, by Lemma 1.17,

$$\liminf_{n\to\infty} \int\int k_\alpha(x-y)\, \mu_n\,(dx)\, \mu_n\,(dy) \geqslant \int\int k_\alpha(x, y)\, \mu\,(dx)\, \mu\,(dy)$$

that is,

$$\liminf_{n\to\infty} \| \mu_n \| \geqslant \| \mu \|. \tag{3.16}$$

Since $\{\| \mu_n \|\}$ is bounded, $\mu \in \mathscr{E}_\alpha$ and, since μ_n is a positive linear functional, so is μ; and so $\mu \in \mathscr{E}_\alpha^+$.

Suppose now that γ is a charge of which the potential is continuous and of compact support. By Theorem 3.11, the set of such charges is strongly dense in $\mathscr{E}_\alpha$.

Now, since we are given $\mu_n \rightsquigarrow \mu$, as n tends to infinity,

$$(\mu_n, \gamma) = \int U^\gamma_\alpha(x)\, \mu_n\,(dx) \to \int U^\gamma_\alpha(x)\, \mu\,(dx) = (\mu, \gamma).$$

Then, given $\kappa \in \mathscr{E}_\alpha$ and $\varepsilon > 0$, we may choose γ so that $\| \kappa - \gamma \| < \varepsilon$ and then choose N so that, for $n > N$,

$$| (\mu_n, \gamma) - (\mu, \gamma) | < \varepsilon.$$

Hence $| (\mu_n, \kappa) - (\mu, \kappa) | \leqslant | (\mu_n, \gamma) - (\mu, \gamma) | + | (\mu_n, \kappa - \gamma) | + | (\mu, \kappa - \gamma) |$ and $| (\mu_n, \kappa - \gamma) | \leqslant \| \mu_n \| \, \| \kappa - \gamma \|$, $| (\mu, \kappa - \gamma) | \leqslant \| \mu \| \, \| \kappa - \gamma \|$, and so for $n > N$, and with $M = \sup_n \| \mu_n \|$,

$$| (\mu_n, \kappa) - (\mu, \kappa) | \leqslant \varepsilon + M\varepsilon + M\varepsilon$$

so that

$$\lim_{n \to \infty} (\mu_n, \kappa) = (\mu, \kappa), \; \kappa \in \mathscr{E}_\alpha.$$

LEMMA 3.13. *Suppose that* $\{\mu_n\}$, $\mu_n \in \mathscr{E}^+_\alpha$ *is a Cauchy sequence, and that* $\mu_n \rightsquigarrow \mu$. *Then* $\mu \in \mathscr{E}^+_\alpha$ *and* $\mu_n \twoheadrightarrow \mu$.

First, since $\{\mu_n\}$ is a Cauchy sequence, $\{\| \mu_n \|\}$ is bounded; so, by Lemma 3.12, $\mu \in \mathscr{E}^+_\alpha$ and $\mu_n \to \mu$. Next,

$$\begin{aligned} \| \mu_n - \mu \|^2 &= (\mu_n - \mu, \mu_n - \mu) = (\mu_n - \mu, \mu_n - \mu_m) + (\mu_n - \mu, \mu_m - \mu) \\ &= (\mu_n - \mu, \mu_n - \mu_m) + (\mu_n - \mu, \mu_m) - (\mu_n - \mu, \mu). \end{aligned}$$

Suppose that $\| \mu_n \| \leqslant M$. Then, given $\varepsilon > 0$, we can find N such that, for $m, n > N$,

$$\| \mu_n - \mu_m \| < \varepsilon^2 / 4M.$$

Consequently,

$$\begin{aligned} | (\mu_n - \mu, \mu_n - \mu_m) | \leqslant \| \mu_n - \mu \| \, \| \mu_n - \mu_m \| &\leqslant (\| \mu_n \| + \| \mu \|) \, \| \mu_n - \mu_m \| \\ &< 2M \frac{\varepsilon^2}{4M} = \tfrac{1}{2}\varepsilon^2. \end{aligned}$$

Finally, given $n > N$, we can find $m > N$ such that

$$| (\mu_n - \mu, \mu_m) - (\mu_n - \mu, \mu) | < \tfrac{1}{2}\varepsilon^2.$$

Thus, for $n > N$, $\| \mu_n - \mu \|^2 < \varepsilon^2$; and so $\mu_n \twoheadrightarrow \mu$.

THEOREM 3.14. $\mathscr{E}^+_\alpha$ *is a complete metric space.*

We must show that, given a Cauchy sequence $\{\mu_n\}$ in $\mathscr{E}_\alpha^+$, there is a $\mu \in \mathscr{E}_\alpha^+$ such that $\mu_n \twoheadrightarrow \mu$. By Lemma 3.13 it is enough to show that there is a measure μ such that $\mu_n \rightsquigarrow \mu$.

Given $\gamma \in \mathscr{E}_\alpha$,

$$|(\mu_n, \gamma) - (\mu_m, \gamma)| = |(\mu_n - \mu_m, \gamma)| \leqslant \| \mu_n - \mu_m \| \, \| \gamma \|$$

so that $(\mu_n \gamma)$ is a Cauchy sequence of real numbers, and hence convergent. Thus

$$\lim_{n\to\infty} \int U_\alpha^\gamma(x) \, \mu_n(dx) \text{ exists for each } \gamma \in \mathscr{E}_\alpha.$$

Let $\phi \in \mathscr{D}(\mathbb{R}^p)$. Then, by Lemma 3.2, there is an infinitely differentiable function ψ such that $\phi = U_\alpha^\psi$ and, by (3.5), $\psi \in \mathscr{E}_\alpha$. Hence $\lim_{n\to\infty} \mu_n(\phi)$ exists for all $\phi \in \mathscr{D}(\mathbb{R}^p)$. Denote this limit by $\mu(\phi)$. It is then clear that μ is a positive linear functional on $\mathscr{D}(\mathbb{R}^p)$. By Theorem 1.30, there is one and only one measure on $\mathbb{R}^p$ which extends μ. Let us denote this measure again by μ. Now

$$\lim_{n\to\infty} \mu_n(\phi) = \mu(\phi), \ \phi \in \mathscr{D}(\mathbb{R}^p).$$

Since $\mathscr{D}(\mathbb{R}^p)$ is dense in $\mathscr{C}_0(\mathbb{R}^p)$, Lemma 1.31 now shows that $\mu_n \rightsquigarrow \mu$.

This completes the proof that $\mathscr{E}_\alpha^+$ is a complete metric space.

§ 3.5. The Conductor Problem

If one charges an electrical conductor with a positive charge, it is an experimental fact that the charge disposes itself on the surface of the conductor so that the electric field within the conductor is zero. This amounts to saying that the electric potential is constant throughout the conductor. The charge is then said to be the equilibrium distribution and the potential the equilibrium potential.

The preceding considerations are purely physical. To give a mathematical treatment one needs, before all else, to show that an equilibrium distribution exists.

Before 1920 a charge distribution was described by giving its density function, and not many people, either physicists or mathematicians, would have thought of doing otherwise. If it is so described it is impossible, except in the simplest situations, to show that an equilibrium distribution exists. G. C. Evans, in 1922, was perhaps the first to point out that a much more natural way of describing a charge distribution was to treat it as a set function—indeed, to regard it as a signed measure, what we have called a charge, in fact.

The classical treatment of the Dirichlet Problem by way of Fredholm integral equations, as exemplified in Kellogg, does in a hidden sense introduce a charge, in so far as a density function defined and integrated on a smooth surface gives rise to a charge supported on the surface; and Kellogg does see the conductor problem as a particular case of the Dirichlet Problem. He shows that it has a unique solution when the conductor consists of a finite number of closed regions each having a smooth surface (pp. 311 *et seq.*). And this, of course, is the conductor problem in $\mathbb{R}^3$ with $\alpha = 2$.

In fact (and here we anticipate the discussion in Chapter 4) the conductor problem in $\mathbb{R}^p$ with $\alpha = 2$ is a special case of a more general Dirichlet Problem. It is a consequence of the discussion in §§7, 8 of Chapter 4 that, given a compact K, we can solve the exterior Dirichlet Problem for $\mathring{K}$, that is, we can find a unique function u harmonic in $\mathbb{R}^p \setminus K$ such that

$$\lim_{x \notin K, x \to \xi} u(x) = 1 \text{ q.e. on } \partial K \text{ and } \lim_{|x| \to \infty} u(x) = 0.$$

Then the function, defined to be 1 in K and at every regular point of ∂K, and defined to be u outside K, can be extended uniquely to a function superharmonic in $\mathbb{R}^p$. Being superharmonic, and because it satisfies the condition of Lemma 4.36,

$$v(x) = \int |x-t|^{2-p} \mu(dt)$$

where μ is the measure associated with v by the Riesz Decomposition. Thus v is the equilibrium potential and $C_p^{-1}\Delta v$ is the equilibrium measure.

All this, however, is limited to the case $\alpha = 2$ and contributes nothing for other values of α. Frostman, in 1935, made a substantial contribution to the problem for other α. From the outset he dealt in α-potentials of charges and utilised the concept of vague convergence as well as the α-energy of a charge. He showed that if D was a finite union of disjoint closed domains, for each of which the boundary satisfies an exterior cone condition, then there is a unique equilibrium distribution μ such that U_μ^α for $\alpha \leqslant 2$ takes the value 1 everywhere in D. He showed further how to extend the result to a compact K of positive capacity.

But he was not able to deal with the case $\alpha > 2$. It needed, as Henri Cartan saw, the recognition that one was in a Hilbert space situation to produce a general method that would deal with all α.

We turn now to give an account of the solution of the Conductor Problem, in which we shall make extensive use of the Hilbert space machinery we have set up.

Let $K \subset \mathbb{R}^p$ be a compact and let $\mathcal{M}_K^*$ denote the set of all measures supported by K. Let $\mathcal{M}_K$ denote the subset of $\mathcal{M}_K^*$ such that $\mu(\mathbb{R}^p) = \mu(K) = 1$. Then

LEMMA 3.15. *$\mathcal{M}_K$ is convex and vaguely compact.*

First, suppose that μ, $\nu \in \mathcal{M}_K$ and that α, $\beta \in \mathbb{R}$ with $\alpha, \beta \geqslant 0$ and $\alpha+\beta = 1$. Then $(\alpha\mu+\beta\nu)(\mathbb{R}^p) = 1$, and $\alpha\mu+\beta\nu$ is supported by K so $\alpha\mu+\beta\nu \in \mathcal{M}_K$. Hence $\mathcal{M}_K$ is convex.

To show that $\mathcal{M}_K$ is vaguely compact, we must show that every sequence $\{\mu_n\}$, $\mu_n \in \mathcal{M}_K$ has a vaguely convergent subsequence.

Let $f \in \mathscr{C}_0(\mathbb{R}^p)$ and choose the least integer k such that supp $f \subset B_{k-1}(0)$. Let ϕ_k be defined by

$$\begin{aligned}\phi_k(x) &= 1 \text{ in } \bar{B}_{k-1}(0)\\ &= k-|x| \text{ in } B_k(0)\setminus\bar{B}_{k-1}(0)\\ &= 0 \text{ in } \mathbb{R}^p\setminus B_k(0).\end{aligned}$$

Given $\varepsilon > 0$, there is, by the Weierstrass approximation theorem (Theorem 1.10), and since the rationals are dense in the reals, a polynomial $p(x)$ with rational coefficients such that

$$|f(x)-p(x)| < \varepsilon \text{ for } x \in \bar{B}_k(0).$$

Thus $|p(x)| < \varepsilon$ in $\bar{B}_k(0)\setminus B_{k-1}(0)$, and so

$$|f(x)-\phi_k(x)\,p(x)| < \varepsilon \text{ for } x \in \mathbb{R}^p.$$

Thus the set

$$S = \{p\phi_k \mid p \text{ a polynomial with rational coefficients}; k = 1, 2, \ldots\}$$

is dense in $\mathscr{C}_0(\mathbb{R}^p)$ and is countable. Denote it by $\{f_m\}$. Now $|\mu_n(f_m)|$ is less than, or equal to, sup $|f_m|$ and so $\{\mu_n(f_m)\}_n$ is a bounded sequence of real numbers for each m. Hence there is a subsequence $\{r_n^1\}$ such that $\{\mu_{r_n^1}(f_1)\}$ converges. Then $\{\mu_{r_n^1}(f_2)\}$ is bounded, and so there is a subsequence $\{r_n^2\}$ of $\{r_n^1\}$ such that $\{\mu_{r_n^2}(f_2)\}$ converges. Proceeding in this way, we can define subsequences $\{r_n^m\}_{n=1,2\ldots}$ for each m, such thas $\{\mu_{r_n^m}(f_m)\}$ converges and $\{r_n^m\}$ is a subsequence of $\{r_n^{m-1}\}$. If we now set $\rho_n = r_n^n$ we see that $\mu_{\rho_n}(f_m)$ converges for every m. Let μ be defined

$$\mu(f_m) = \lim_{n\to\infty} \mu_{\rho_n}(f_m).$$

Let M be the set of all real linear combinations of S. Then M is a dense linear subspace of $\mathscr{C}_0(\mathbb{R}^p)$. Extend μ from S to M by linearity. Then μ is a positive linear functional on M. Clearly, the set M is such that, given $f \in \mathscr{C}_0(\mathbb{R}^p)$, we can find $\{\phi_n\}$, $\phi_n \in M$ such that $\phi_n \to f$ uniformly and supp $\phi_n \subset K$, K a compact independent of n. Hence $\{\phi_n\}$ satisfies (*a*) and (*b*) in the proof of Theorem 1.30 which thus applies to this situation. Thus μ can be extended in one and only one way to a measure. Then Lemma 1.31 shows that $\mu_{\rho_n} \rightsquigarrow \mu$. Since $\mu_{\rho_n} \in \mathcal{M}_K$, μ is also a member of $\mathcal{M}_K$, and hence $\mathcal{M}_K$ is vaguely compact.

THEOREM 3.16. *There is a measure $\lambda_K \in \mathscr{M}_K$ such that*

$$\| \lambda_K \| \leqslant \| \mu \| \text{ for all } \mu \in \mathscr{M}_K.$$

Let

$$W_\alpha(K) = \inf \{\| \mu \|^2 \mid \mu \in \mathscr{M}_K\}.$$

If $W_\alpha(K) = \infty$, the theorem is vacuous; so we may suppose that $W_\alpha(K)$ is finite. There is then a sequence $\{\mu_n\}$, $\mu_n \in \mathscr{M}_K$ such that $\| \mu_n \|^2 \to W_\alpha(K)$. By Lemma 3.15, $\{\mu_n\}$ has a vaguely convergent subsequence, and we may therefore suppose, without loss of generality, that $\{\mu_n\}$ is convergent and that $\mu_n \rightsquigarrow \lambda_K$. Also, by (3.16),

$$\lim \| \mu_n \|^2 \geqslant \| \lambda_K \|^2 \text{ and so } \| \lambda_K \|^2 = W_\alpha(K) \text{ and } \lambda_K \in \mathscr{M}_K.$$

The sequence $\{\mu_n\}$ introduced in the proof is said to be a *minimising sequence* and λ_K is said to be a minimising measure.

THEOREM 3.17. *The minimising measure is unique and any minimising sequence converges strongly to λ_K.*

For suppose that λ_K is a minimising measure and $\{\mu_n\}$ is a minimising sequence. Then $\mu_n \in \mathscr{E}_\alpha^+$ and a subsequence of $\{\mu_n\}$ converges vaguely to λ_K.

Since $\mathscr{M}_K$ is convex, $\frac{1}{2}(\mu_n + \mu_m) \in \mathscr{M}_K$, and so

$$\sqrt{W_\alpha(K)} \leqslant \| \tfrac{1}{2}(\mu_n + \mu_m) \| \leqslant \tfrac{1}{2} \| \mu_n \| + \tfrac{1}{2} \| \mu_m \|$$

and since $\| \mu_n \| \to \sqrt{W_\alpha(K)}$ we have $\| \frac{1}{2}(\mu_n + \mu_m) \| \to W_\alpha(K)$ as $m, n \to \infty$.

Now

$$\| \mu_n - \mu_m \|^2 = 2 \| \mu_n \|^2 + 2 \| \mu_m \|^2 - 4 \| \tfrac{1}{2}(\mu_n + \mu_m) \|^2$$

and so $\| \mu_n - \mu_m \| \to 0$ as $m, n \to \infty$.

Thus $\{\mu_n\}$ is a Cauchy sequence in $\mathscr{E}_\alpha^+$ and has a vaguely convergent subsequence. This subsequence, being Cauchy, is, by Lemma 3.13, strongly convergent to λ_K. Consequently, $\{\mu_n\}$ itself converges strongly to λ_K.

Suppose that λ and λ' are both minimising measures. Then $\lambda, \lambda' \in \mathscr{M}_K$ and so $\frac{1}{2}(\lambda + \lambda') \in \mathscr{M}_K$. Hence

$$\sqrt{W_\alpha(K)} \leqslant \| \tfrac{1}{2}(\lambda + \lambda') \| \leqslant \tfrac{1}{2} \| \lambda \| + \tfrac{1}{2} \| \lambda' \| = \sqrt{W_\alpha(K)}$$

and so $\| \frac{1}{2}(\lambda + \lambda') \| = \sqrt{W_\alpha(K)}$ and

$$\| \lambda - \lambda' \|^2 = 2 \| \lambda \|^2 + 2 \| \lambda' \|^2 - 4 \| \tfrac{1}{2}(\lambda + \lambda') \|^2 = 0.$$

Consequently $\lambda = \lambda'$ and the minimising measure is unique.

We define the **α-capacity** of a compact K, cap K, by

$$\text{cap } K = (W_\alpha(K))^{-1}.$$

At a later stage we shall be able to show that this definition is consistent with that adopted in electrostatics.

If $W_\alpha(K) = \infty$ we say that cap $K = 0$ and that K is of **zero α-capacity.** We note then that if cap $K = 0$ then $\mathscr{M}_K$ is void. This gives

THEOREM 3.18. *cap $K = 0$ if, and only if, $\mu(K) = 0$ for all $\mu \in \mathscr{E}_\alpha^+$.*

For if $\mu(K) = 0$ for all $\mu \in \mathscr{E}_\alpha^+$, then there is no $\mu \in \mathscr{M}_K$ which is of finite energy, and so $W_\alpha(K) = \infty$. Hence cap $K = 0$. If, conversely, cap $K = 0$ and $\mu \in \mathscr{E}_\alpha^+$, then its restriction μ_K to K also belongs to $\mathscr{E}_\alpha^+$ since $\| \mu_K \| \leqslant \| \mu \|$. But $W_\alpha(K) = \infty$. Therefore, if $\mu(K) \neq 0$, then the measure $\mu_K/\mu(K) \in \mathscr{M}_K$ and is of finite α-energy and so $W_\alpha(K) < \infty$—a contradiction.

It follows from this that if cap $K = 0$, then the Lebesque measure of K is zero, since the restriction of Lebesque measure to the compact K is a measure of finite energy.

LEMMA 3.19. *If cap $K = 0$ and $\mu \in \mathscr{M}_K$ then sup $\{U_\alpha^\mu(x) \mid x \in \operatorname{supp} \mu\} = +\infty$.*

For if not, then $\| \mu \|^2 = \int_{\operatorname{supp} \mu} U_\alpha^\mu(x)\, \mu\,(dx) < \infty$ and then cap $K > 0$.

Thus far, we have defined capacity only for compact sets; we now need the concept of a set (not necessarily compact) of zero interior capacity. We shall say that a set E is of **zero interior α-capacity** if, for every compact $K \subset E$,

$$\text{cap } K = 0.$$

We say that a property is true **nearly everywhere** (which we abbreviate to n.e.), if it is true everywhere except possibly in a set of zero interior capacity. In what follows we shall omit to state that it is α-capacity, α-nearly everywhere, etc., that is being spoken of, so long as there is no danger of ambiguity.

We may note that Lemma 3.18 has the analogue

E has zero interior capacity if, and only if, $\mu(E) = 0$ for all $\mu \in \mathscr{E}_\alpha^+$, and omit the easy proof.

THEOREM 3.20. *Let K be a compact and λ its minimising measure. Then*

(*a*) $U_\alpha^\lambda(x) \geqslant W_\alpha(K) = \| \lambda \|^2$ *n.e. in K;*

(*b*) $U_\alpha^\lambda(x) \leqslant W_\alpha(K)$ *when $x \in \operatorname{supp} \lambda$;*

(*c*) $U_\alpha^\lambda(x) = W_\alpha(K)$ *n.e. in supp λ.*

Suppose that (*a*) is false. Then there is a compact $K_0 \subset K$ with cap $K_0 > 0$ such that $U_\alpha^\lambda(x) < \| \lambda \|^2$ for $x \in K_0$. Suppose $\nu \in \mathscr{M}_K$ and of finite energy. Then

$$(\nu, \lambda) = \int U_\alpha^\lambda(x)\, \nu\,(dx) < \| \lambda \|^2 \tag{3.17}$$

But $a\nu+(1-a)\lambda \in \mathscr{M}_K$ and is of finite energy for $a \in \mathbb{R}$, $0 \leqslant a \leqslant 1$, so

$$\| a\nu+(1-a)\lambda \| \geqslant \| \lambda \|,$$

whence

$$a^2 \| \nu \|^2+2a(1-a)(\nu, \lambda)+(1-a)^2 \| \lambda \|^2 \geqslant \| \lambda \|^2$$

which gives, for $a \neq 0$,

$$2[(\nu, \lambda)-\| \lambda \|^2] \geqslant a[2(\nu, \lambda)-\| \lambda \|^2-\| \nu \|^2].$$

Letting $a \to 0$ we have $(\nu, \lambda) \geqslant \| \lambda \|^2$; which contradicts (3.17). Thus (*a*) is proved.

As to (*b*), suppose, if possible, that there is $x_0 \in \operatorname{supp} \lambda$ such that $U_\alpha^\lambda(x_0) > W_\alpha(K)$. Since U_α^λ is l.s.c. there is a neighbourhood V of x_0 such that, for $x \in V$,

$$U_\alpha^\lambda(x) > W_\alpha(K)$$

and, for this V, $\lambda(V) > 0$ (for, if not, $V \cap \operatorname{supp} \lambda = \phi$). Hence

$$\begin{aligned} \| \lambda \|^2 &= \int_V U_\alpha^\lambda(x)\, \lambda(dx)+\int_{\operatorname{supp} \lambda \setminus V} U_\alpha^\lambda(x)\, \lambda(dx) \\ &> W_\alpha(K)\, \lambda(V)+W_\alpha(K)(1-\lambda(V)) = W_\alpha(K) \end{aligned}$$

where we use (*a*) to estimate the second integral. But $\| \lambda \|^2 = W_\alpha(K)$ and this contradiction then proves (*b*).

Since $\operatorname{supp} \lambda \subset K$ (*c*) now follows from (*a*) and (*b*).

When $\alpha = 2$, all that we have said so far will concern the Newtonian potential and, indeed, in that case much more is true as we shall see presently. In fact, as we change from $\alpha > 2$ to $\alpha \leqslant 2$ there is a radical change in the behaviour of the equilibrium potential. To show this we first need a number of preparatory lemmas.

LEMMA 3.21. *Let $p \geqslant 2$ and $R > r > 0$. Let $0 < \alpha < p$. Then*

$$\int_0^\pi \frac{(R \cos \theta-r) \sin^{p-2} \theta\, d\theta}{(R^2-2Rr \cos \theta+r^2)^{\frac{1}{2}(p-\alpha+2)}}$$

is > 0 when $\alpha < 2$, $= 0$ when $\alpha = 2$, and < 0 when $\alpha > 2$.

Set $\cos \psi = (R \cos \theta-r)\rho^{-1}$, $\sin \psi = R \sin \theta\, \rho^{-1}$ where

$$\rho = \sqrt{(R^2-2Rr \cos \theta+r^2)}.$$

Then $\rho = -r \cos \psi+\sqrt{(R^2-r^2 \sin^2 \psi)}$ and $d\theta = \rho(R^2-r^2 \sin^2 \psi)^{-\frac{1}{2}} d\psi$ and the integral becomes

$$R^{2-p} \int_0^\pi \rho^{\alpha-2} \sin^{p-2} \psi \cos \psi\, (R^2-r^2 \sin^2 \psi)^{-\frac{1}{2}}\, d\psi$$

which may be rewritten as

$$R^{2-p} \int_0^{\frac{1}{2}\pi} \{[\sqrt{(R^2-r^2 \sin^2 \psi)}-r \cos \psi]^{\alpha-2}-[\sqrt{(R^2-r^2 \sin^2 \psi)} +r \cos \psi]^{\alpha-2}\} \sin^{p-2} \psi \cos \psi \, (R^2-r^2 \sin^2 \psi)^{-\frac{1}{2}} \, d\psi.$$

Since the integrand is continuous, is strictly positive in $(0, \frac{1}{2}\pi)$ for $\alpha < 2$, strictly negative for $\alpha > 2$, and vanishes for $\alpha = 2$, the lemma follows.

Lemma 3.22.

$$\int | x-t |^{\alpha-p} \sigma_R^0 (dt) \begin{cases} < R^{\alpha-p} \text{ for } 0 < | x | \leqslant R \\ < | x |^{\alpha-p} \text{ for } | x | > R \end{cases} \quad \text{when } \alpha > 2,$$

and

$$\int | x-t |^{\alpha-p} \sigma_R^0 (dt) \begin{cases} > R^{\alpha-p} \text{ for } 0 < | x | \leqslant R \\ > | x |^{\alpha-p} \text{ for } | x | > R \end{cases} \quad \text{when } \alpha < 2.$$

When $\alpha = 2$ we have equality in every case.

(Note that the case $\alpha = 2$ is Lemma 2.15 when $p \geqslant 3$.)

Let $| x | = r$, and choose coordinates so that $x = (r, 0, \ldots, 0)$. Transform to spherical polar coordinates. Then

$$| x-t |^2 = R^2-2Rr \cos \theta_1+r^2$$

and the integral in the lemma becomes a constant multiple of

$$\int_0^{\pi} (R^2-2Rr \cos \theta+r^2)^{\frac{1}{2}(\alpha-p)} \sin^{p-2} \theta \, d\theta.$$

Suppose now that $r < R$. The derivative with respect to r of this last integral is

$$(\alpha-p) \int_0^{\pi} (R^2-2Rr \cos \theta+r^2)^{\frac{1}{2}(\alpha-p)-1} (r-R \cos \theta) \sin^{p-2} \theta \, d\theta,$$

which, by Lemma 3.21, is negative for $\alpha < 2$ and positive for $\alpha > 2$. Hence the integral in the lemma is, as a function of r, strictly decreasing for $\alpha < 2$ and strictly increasing for $\alpha > 2$ and takes the value $R^{\alpha-p}$ when $r = 0$. This deals with the case $0 < | x | < R$.

Suppose next that $r > R$, and consider

$$r^{p-\alpha} \int | x-t |^{\alpha-p} \sigma_R^0 (dt). \tag{3.18}$$

It is a constant multiple of

$$r^{p-\alpha} \int_0^{\pi} (R^2-2Rr \cos \theta+r^2)^{\frac{1}{2}(\alpha-p)} \sin^{p-2} \theta \, d\theta$$

and the derivative of this with respect to r is

$$(p-\alpha)r^{p-\alpha-1}\int_0^{\pi}\frac{R(R-r\cos\theta)\sin^{p-2}\theta}{(R^2-2Rr\cos\theta+r^2)^{\frac{1}{2}(\alpha-p)-1}}\,d\theta$$

which, by Lemma 3.21, noting that the roles of r and R are here interchanged, is strictly negative when $\alpha < 2$ and strictly positive when $\alpha > 2$ and so (3.18) is strictly decreasing when $\alpha < 2$ and strictly increasing when $\alpha > 2$. Finally, as $r \to \infty$, $r^{p-\alpha}\,|\,x-t\,|^{\alpha-p}$ tends to 1 uniformly, and so (3.18) tends to 1. This deals with the case $|\,x\,| > R$.

When $\alpha = 2$ the derivatives are identically zero, the functions concerned are constant and the final result follows. Furthermore, the case $p = 2$ in Lemma 2.15 follows if we write $\log r$ in place of $r^{\alpha-p}$.

LEMMA 3.23. *For $\alpha > 2$, $U_\alpha^\mu(x)$ is strictly superharmonic, that is,*

$$\int U_\alpha^\mu(x)\,\sigma_R^a\,(dx) < U_\alpha^\mu(a)$$

and $U_\alpha^\mu(x)$ is l.s.c.

Let

$$V_\alpha^\mu(x) = \int |\,x-t\,|^{\alpha-p}\,\mu(dt).$$

Then

$$\begin{aligned}\int V_\alpha^\mu(x)\,\sigma_R^a(dx) &= \int V_\alpha^\mu(x+a)\,\sigma_R^0(dx)\\ &= \int\int |\,x+a-t\,|^{\alpha-p}\,\sigma_R^0(dx)\,\mu(dt).\end{aligned} \tag{3.19}$$

Now, by Lemma 3.22, we have, in fact,

$$\int |\,z-t\,|^{\alpha-p}\,\sigma_R^0\,(dt) < |\,z\,|^{\alpha-p} \text{ for all } z,$$

and so the double integral is strictly less than

$$\int |\,t-a\,|^{\alpha-p}\,\mu\,(dt) = V_\alpha^\mu(a),$$

which, since V_α^μ is a constant multiple of U_α^μ, gives the 'super-mean' inequality.

Also

$$\begin{aligned}\liminf_{x\to a} V_\alpha^\mu(x) &\geqslant \int \liminf_{x\to a} |\,x-t\,|^{\alpha-p}\,\mu\,(dt)\\ &= \int |\,t-a\,|^{\alpha-p}\,\mu\,(dt) = V_\alpha^\mu(a),\end{aligned}$$

showing that V_α^μ, and hence U_μ^μ is l.s.c.

The case $\alpha < 2$ does not lead to so pleasing a result. We then have

LEMMA 3.24. *For $\alpha < 2$, $U_\alpha^\mu(x)$ is strictly subharmonic in $\mathbb{R}^p \setminus \operatorname{supp}\mu$.*

Choose $R > 0$ so that $\bar{B}_R(a) \cap \operatorname{supp} \mu = \phi$. Now, if $t \in \operatorname{supp} \mu$, then $|t-a| > R$ and so, by Lemma 3.22,

$$\int |x+a-t|^{\alpha-p} \sigma_R^0(dx) > |t-a|^{\alpha-p}.$$

Consequently (3.19) is strictly greater than $V_\mu^\mu(a)$, and we thus have the required sub-mean inequality.

Suppose now that $a \notin \operatorname{supp} \mu$. Then there is an $R > 0$ such that $B_{2R}(a) \cap \operatorname{supp} \mu = \phi$. Consequently, if $x \in B_R(a)$ and $t \in \operatorname{supp} \mu$, then $|x-t|^{\alpha-p} < R^{\alpha-p}$ and so, by Theorem 1.23,

$$\lim_{x \to a} V_\alpha^\mu(x) = V_\alpha^\mu(a).$$

Thus U_α^μ is continuous in $\mathbb{R}^p \setminus \operatorname{supp} \mu$ and the theorem follows.

By making the necessary changes in the arguments in Lemma 3.23 and Lemma 3.24 we may show that

LEMMA 3.25. *When* $\alpha = 2$, $U_\alpha^\mu(x)$ *is superharmonic and is harmonic in* $\mathbb{R}^p \setminus \operatorname{supp} \mu$.

Lemma 3.24 has nothing to say about the behaviour of U_α^μ in supp μ. The next lemma shows that there is a sort of generalised supermean relationship.

LEMMA 3.26. *Suppose that* $\alpha < 2$. *Then there is a constant* A_α, *depending only on* α, *such that*

$$\mathscr{B}_R^a(U_\alpha^\mu) \leqslant A_\alpha U_\alpha^\mu(a).$$

We have, after inverting the order of integration,

$$\mathscr{B}_R^a(V_\alpha^\mu) = \int |t-a|^{\alpha-p} \left\{\frac{\Gamma(\frac{1}{2}p+1)}{\pi^{\frac{1}{2}p} R^p} \int_{|x| \leqslant R} |x+a-t|^{\alpha-p} |t-a|^{p-\alpha} dx\right\} \mu(dt).$$

The expression in curly brackets tends, as $|t| \to \infty$, to 1 and, when $t = a$ has the value 0. Furthermore, making the substitution $x = |t-a|u$, we obtain for it

$$\frac{\Gamma(\frac{1}{2}p+1)}{\pi^{\frac{1}{2}p}} \left(\frac{|t-a|}{R}\right)^p \int_{|u| \leqslant R/|t-a|} |u-w|^{\alpha-p} du, \; w = (t-a)/|t-a|.$$

This last integral is independent of the unit vector w and so the expression in curly brackets depends only on $(|t-a|/R)^p$, has the value 0 when $t = a$, and tends to 1 as $|t| \to \infty$, and is, furthermore, continuous in this variable. It therefore attains a maximum A_α. Consequently, since U_α^μ is a constant multiple of V_α^μ,

$$\mathscr{B}_R^a(U_\alpha^\mu) \leqslant A_\alpha \int |t-a|^{\alpha-p} \mu(dt) = A_\alpha U_\alpha^\mu(a).$$

We may now turn to the three cases $\alpha > 2, \alpha < 2$ and $\alpha = 2$.

THEOREM 3.27. *Suppose that* $2 < \alpha < p$, *that* K *is a compact and that* λ *is the minimising measure for* K. *Then*

(*a*) $U_\alpha^\lambda(x) > W_\alpha(K)$ *when* $x \in \mathring{K}$;

(*b*) $U_\alpha^\lambda(x) \leqslant 2^{p-\alpha}\, W_\alpha(K)$ *everywhere in* $\mathbb{R}^p$;

(*c*) $U_\alpha^\lambda(x) = W_\alpha(K)$ *n.e. in supp* $\lambda \subset \partial K$.

Let $x \in \mathring{K}$. Then there is an $R > 0$ such that $B_R(a) \subset K$ and, since U_α^λ is strictly superharmonic, $U_\alpha^\lambda(x) > \mathscr{B}_R^x(U_\alpha^\lambda)$ (using (1.9) in Chapter 1). Since $U_\alpha^\lambda(t) \geqslant W_\alpha(K)$ n.e. in K and hence a.e. in K and so a.e. in $B_R(a)$ we have $U_\alpha^\lambda(x) > W_\alpha(K)$, giving (*a*).

Suppose that $x \notin \operatorname{supp} \lambda$. Then, since supp λ is closed, we can find $b \in \operatorname{supp} \lambda$ such that, for $t \in \operatorname{supp} \lambda$, $| b-x | \leqslant | t-x |$. Then

$$2\,| t-x | \geqslant | t-x |+| b-x | \geqslant | t-b |,$$

and so

$$U_\alpha^\lambda(x) = C\int | t-x |^{\alpha-p}\,\lambda(dt) \leqslant 2^{p-\alpha}\, C\int | t-b |^{\alpha-p}\,\lambda(dt) = 2^{p-\alpha}\, U_\alpha^\lambda(b).$$

Now $b \in \operatorname{supp} \lambda$ so, by Theorem 3.20(*b*), $U_\alpha^\lambda(x) \leqslant 2^{p-\alpha}\, W_\alpha(K)$ when $x \notin \operatorname{supp} \lambda$. Again using Theorem 3.20(*b*) we now have (*b*).

Finally (*a*) and Theorem 3.20(*b*) show that supp $\lambda \cup \mathring{K} = \phi$ and so supp $\lambda \subset \partial K$. This and Theorem 3.20(*c*) give (*c*).

THEOREM 3.28. *Suppose that* $0 < \alpha < 2$ *and that* λ *and* K *are as in Theorem* 3.27. *Then*

(*a*) $U_\alpha^\lambda(x) = W_\alpha(K)$ *n.e. in* K;

(*b*) *supp* $\lambda \supset \mathring{K}$;

(*c*) $U_\alpha^\lambda(x) \leqslant W_\alpha(K)$ *everywhere in* $\mathbb{R}^p$;

(*d*) $U_\alpha^\lambda(x) = W_\alpha(K)$ *in* $\mathring{K}$.

By Lemma 3.24, and since $U_\alpha^\lambda(x) \to 0$ as $| x | \to \infty$ and cannot take its maximum outside supp λ, we must have

$$U_\alpha^\lambda(x) \leqslant W_\alpha(K) \text{ everywhere in } \mathbb{R}^p$$

giving (*c*) and, furthermore,

$$U_\alpha^\lambda(x) < W_\alpha(K) \text{ outside supp } \lambda. \tag{3.20}$$

By Theorem 3.20(*a*), $U_\alpha^\lambda(x) = W_\alpha(K)$ n.e. in K giving (*a*).

Let $x \in \mathring{K}$ and let $B_n = B_{1/n}(x)$. Now, by (*c*), $U_\alpha^\lambda(x)$ is bounded so

$$\int_{B_1} | t-x |^{\alpha-p}\, \lambda\,(dt) = \sum_{n=1}^{\infty}\ \int_{B_{n-1}\setminus \bar{B}_n} | t-x |^{\alpha-p}\, \lambda\,(dt) < \infty$$

and hence, given $\varepsilon > 0$, there is N such that the sum from N to ∞ of the above series is less than ε, that is,

$$\int_{B_N} |t-x|^{\alpha-p} \lambda(dt) < \varepsilon.$$

Thus, given $\varepsilon > 0$, we can find a ball $B_R(x)$ such that

$$\int_{B_R(x)} |t-x|^{\alpha-p} \lambda(dt) < \varepsilon/2A_\alpha C$$

A_α being the constant in Lemma 3.26 and C that in Lemma 3.27.

Let λ_1 be the restriction of λ to $B_R(x)$, and λ_2 the restriction of λ to $\mathbb{R}^p \setminus B_R(x)$. Then

$$U_\alpha^{\lambda_1}(x) < \varepsilon/2A_\alpha.$$

Also $U_\alpha^{\lambda_2}$ is continuous in $B_R(x)$ since supp $\lambda_2 \cap B_R(x) = \phi$ and so we can find a ball $B_r(x) \subset B_R(x)$ such that, for $t \in B_r(x)$,

$$U_\alpha^{\lambda_2}(t) < U_\alpha^{\lambda_2}(x) + \tfrac{1}{2}\varepsilon.$$

Now, by (a)

$$\begin{aligned} W_\alpha(K) &\leqslant \mathscr{B}_r^x(U_\alpha^\lambda) = \mathscr{B}_r^x(U_\alpha^{\lambda_1}) + \mathscr{B}_r^x(U_\alpha^{\lambda_2}) < \mathscr{B}_r^x(U_\alpha^{\lambda_1}) + U_\alpha^{\lambda_2}(x) + \tfrac{1}{2}\varepsilon \\ &\leqslant A_\alpha\, U_\alpha^{\lambda_1}(x) + U_\alpha^{\lambda_2}(x) + \tfrac{1}{2}\varepsilon < U_\alpha^{\lambda_2}(x) + \varepsilon. \end{aligned}$$

Since ε is arbitrary we then have

$$U_\alpha^\lambda(x) \geqslant U_\alpha^{\lambda_2}(x) \geqslant W_\alpha(K)$$

which, with (c), gives (d)

(d), together with (3.20), gives (b).

We note then that there is a sharp distinction between the cases $\alpha > 2$ and $\alpha < 2$. In the former the maximising measure sits on the outer skin of the conductor and the associated potential remains constant only on the skin and increases as we pass to the interior of the conductor, while in the latter the measure fills the whole of the conductor but the associated potential is constant (effectively) everywhere on the conductor.

The case $\alpha = 2$ has, as it were, the best of both worlds, as the following theorem shows.

THEOREM 3.29. *Suppose that* $\alpha = 2$ *and* λ *and* K *are as before. Then*

(a) $U_2^\lambda(x) = W_2(K)$ *n.e. in* K *and everywhere in* $\mathring{K}$;

(b) $U_2^\lambda(x) \leqslant W_2(K)$ *everywhere in* $\mathbb{R}^p$;

(c) *supp* $\lambda \subset \partial K$.

The argument in Lemma 3.26 with $A_\alpha = 1$ (since U_2^λ is superharmonic) is valid here to prove equality in $\mathring{K}$. Then (*b*) follows since U_2^λ is harmonic outside supp λ, and this, with Theorem 3.20(*a*) gives (*a*).

We now turn to (*c*). First, when $|z| < R$,

$$\int |z-t|^{2-p}\, \beta_R^0(dt) < |z|^{2-p}.$$

For, by the formula (1.9),

$$\int |z-t|^{2-p}\, \beta_R^0(dt) = \frac{p}{R^p}\Big(\int_0^{|z|} + \int_{|z|}^{R}\Big)\, \rho^{p-1}\, \Big(\int |z-t|^{2-p}\, \sigma_\rho^0(dt)\Big)\, d\rho$$

and, by Lemma 3.22, the inner integral has the value $|z|^{2-p}$ when $|z| > \rho$ and ρ^{2-p} when $|z| \leqslant \rho$. Consequently, the right-hand side becomes

$$\frac{p}{R^p}\int_0^{|z|} \rho^{p-1}\, |z|^{2-p}\, d\rho + \frac{p}{R^p}\int_{|z|}^{R} \rho\, d\rho$$

$$= \frac{p}{2}\, R^{2-p} + \frac{2-p}{2}\, |z|^2\, R^{-p} \leqslant R^{2-p}$$

and, since $|z| < R$, this last is less than $|z|^{2-p}$.

Next, when $|z| \geqslant R$, the consequent change in the above formula shows that

$$\int |z-t|^{2-p}\, \beta_R^0(dt) = |z|^{2-p}.$$

Now suppose that supp $\lambda \cap \mathring{K} \neq \phi$. Then there is a ball $B_R(a) \subset K$ such that $\lambda(B_R(a)) > 0$. Since $U_2^\lambda(x) = W_2(K)$ a.e. in K,

$$W_\alpha(K) = \mathscr{B}_R^a(U_2^\lambda) = C\Big(\int_{|t-a|<R} + \int_{|t-a|\geqslant R}\Big)\{\int |x+a-t|^{2-p}\, \beta_R^0(dx)\}\, \lambda(dt)$$

$$< C\int |t-a|^{2-p}\, \lambda(dt) = U_2^\lambda(a).$$

Thus $U_2^\lambda(a) > W_2(K)$; which contradicts (*b*). Thus (*c*) follows.

§ 3.6. Capacity

Let K be a compact and let λ be its minimising measure. Define the measure γ by

$$\gamma = (W_\alpha(K))^{-1}\, \lambda.$$

Then $U_\alpha^\gamma(x) = 1$ n.e. in supp λ, and $\gamma(\mathbb{R}^p) = (W_\alpha(K))^{-1}$.

When $\alpha \leqslant 2$, then $U_\alpha^\gamma(x) = 1$ n.e. in K; there is equality in $\mathring{K}$ and $U_\alpha^\gamma(x) \leqslant 1$ everywhere. When $\alpha > 2$, $U_\alpha^\gamma(x) > 1$ in K and $U_\alpha^\gamma(x) < 2^{p-\alpha}$ everywhere.

The measure γ is said to be the **equilibrium measure** for K, and U_α^γ is said to be the **equilibrium potential.** We note that

$$\|\gamma\|^2 = (\text{cap } K)^2\, \|\lambda\|^2 = \text{cap } K.$$

Consider the following extremum problems. To find:

$A = \min \{[\sup U_\alpha^\mu(x) \mid x \in \text{supp}\,\mu] \mid \mu(\mathbb{R}^p) = \text{cap}\,K \text{ and supp}\,\mu \subset K\}$
and, setting $f(\mu) = \mu(\mathbb{R}^p)/[\sup U_\alpha^\mu(x) \mid x \in \text{supp}\,\mu]$

$B = \max \{f(\mu) \mid \text{supp}\,\mu \subset K\}$

$C = \max \{f(\mu) \mid \text{supp}\,\mu \subset K \text{ and } \mu(\mathbb{R}^p) = \text{cap}\,K\}$

$D = \max \{f(\mu) \mid \text{supp}\,\mu \subset K \text{ and } [\sup U_\alpha^\mu(x) \mid x \in \text{supp}\,\mu] = 1\}$

Let γ be the equilibrium measure for K. Then γ is one of the measures in question in A and $U_\alpha^\gamma(x) \leqslant 1$ in supp γ; so $A \leqslant 1$.

Also, if $\mu(\mathbb{R}^p) = \text{cap}\,K$ and $[\sup U_\alpha^\mu(x) \mid x \in \text{supp}\,\mu] \leqslant 1$, then

$$\| \mu \|^2 = \int U_\alpha^\mu(x)\, \mu\,(dx) \leqslant \int \mu\,(dx) = \text{cap}\,K = \| \gamma \|^2$$

so that $\nu = \mu/\text{cap}\,K \in \mathcal{M}_K$ and $\| \nu \|^2 \leqslant \| \lambda \|^2$. Since λ is the unique minimising measure, we have $\nu = \lambda$; and so $\mu = \gamma$. Thus

$A = 1$ *and this value is attained when, and only when,* $\mu = \gamma$.

Next, $f(\mu) = f(a\mu)$ when $a > 0$, and so the extra restrictions for C and D do not diminish the range of $f(\mu)$. Consequently $B = C = D$. In C, $f(\mu) = \text{cap}\,K/[\sup U_\alpha^\mu]$, and so C is attained when, and only when, A is attained. Since $A = 1$, $C = \text{cap}\,K$ and is attained only when $\mu = \gamma$. Furthermore, in D the maximum will not be changed if we replace '$=$' in the condition by '$\leqslant$' and $f(\mu) = \mu(\mathbb{R}^p)$. Consequently

$$\text{cap}\,K = \max \{\mu(\mathbb{R}^p) \mid \text{supp}\,\mu \subset K \text{ and } [\sup U_\alpha^\mu(x) \mid x \in \text{supp}\,\mu] \leqslant 1\}.$$

We are thus led to a formulation which served de la Vallée Poussin as a definition of capacity. When $\alpha \leqslant 2$, it takes the even more pleasing form

$$\text{cap}\,K = \max \{\mu(\mathbb{R}^p) \mid \text{supp}\,\mu \subset K \text{ and } U_\alpha^\mu(x) \leqslant 1 \text{ in } \mathbb{R}^p\}$$

and then, also, the measure γ for which the maximum is attained is such that $U_\alpha^\gamma(x) = 1$ n.e. in K.

The connection between the rather sophisticated mathematical considerations here and the usual definition of capacity in elementary electrostatics now becomes clear, if we remind ourselves that the charge Q there is here represented by $\gamma(K) = \gamma(\mathbb{R}^p)$ and the (constant) potential V on the conductor is here $U_\alpha^\lambda(x)$.

The de la Vallée Poussin formulation is well suited for deriving some properties of capacity qua set function.

LEMMA 3.30. (1) *Let K be a compact and let, for $a \in \mathbb{R}^p$, $a+K = \{a+x \mid x \in K\}$. Then*

$$\textit{cap}\,(a+K) = \textit{cap}\,K.$$

(2) *Let, for $b \in \mathbb{R}$ and $b > 0$, $bK = \{bx \mid x \in K\}$. Then*

$$\textit{cap}\,(bK) = b^{p-\alpha}\,\textit{cap}\,K.$$

Suppose that γ is the equilibrium measure for K. Define γ_a by

$$\gamma_a(a+M) = \gamma(M).$$

Then $\int f(t)\,\lambda_a(dt) = \int f(t+a)\,\lambda(dt)$. For $\{a+M_n\}$ is a partition of $\mathbb{R}^p$ if, and only if, $\{M_n\}$ is. Consequently,

$$\sup \sum \inf_{t \in a+M_n} f(t)\,\gamma_a(a+M_n) = \sup \sum \inf_{t \in M_n} f(t+a)\,\gamma(M_n)$$

and the result follows. Furthermore

$$U_\alpha^{\gamma_a}(x) = \int |x-t|^{\alpha-p}\,\gamma_a(dt) = \int |x-t-a|^{\alpha-p}\,\gamma(dt) = U_\alpha^\gamma(x-a)$$

and so γ_a is the equilibrium measure for $a+K$. Thus

$$\operatorname{cap} K = \gamma(K) = \gamma_a(a+K) = \operatorname{cap}(a+K).$$

Next define the measure v by $v(bM) = \gamma(M)$. Then

$$\int f(t)\,v(dt) = \int f(bt)\,\gamma(dt).$$

For $\{bM_n\}$ is a partition of $\mathbb{R}^p$ if, and only if, $\{M_n\}$ is. Thus

$$\sup \sum \inf_{t \in bM_n} f(t)\,v(bM_n) = \sup \sum \inf_{t \in M_n} f(bt)\,\gamma(M_n)$$

giving the result.

Define γ_b by $\gamma_b(bM) = b^{p-\alpha}\gamma(M)$. Then

$$\begin{aligned} U_\alpha^{\gamma_b}(bx) &= b^{p-\alpha}\,U_\alpha^v(bx) = b^{p-\alpha}\int |bx-t|^{\alpha-p}\,v(dt) \\ &= b^{p-\alpha}\int |bx-bt|^{\alpha-p}\,\gamma(dt) = U_\alpha^\gamma(x) \end{aligned}$$

and so γ_b is the equilibrium measure for bK. Thus

$$\operatorname{cap}(bK) = \gamma_b(bK) = b^{p-\alpha}\gamma(K) = b^{p-\alpha}\operatorname{cap} K.$$

Next, cap is an increasing set function, that is,

LEMMA 3.31. *If $K_1 \subset K_2$ then cap $K_1 \leqslant$ cap K_2.*

If supp $\mu \subset K_1$ then supp $\mu \subset K_2$, and so the set of admissible measures for computing cap K_2 contains that for cap K_1, and the lemma follows.

Next, cap is sub-additive, that is,

LEMMA 3.32. *If K_1 and K_2 are compacts*

$$\operatorname{cap}(K_1 \cup K_2) \leqslant \operatorname{cap} K_1 + \operatorname{cap} K_2.$$

Let

$$\mathscr{S}_K = \{\mu \mid \operatorname{supp}\mu \subset K \text{ and } \sup\,[U_\alpha^\mu(x) \mid x \in \operatorname{supp}\mu] \leqslant 1\}.$$

Let γ be the equilibrium measure for $K_1 \cup K_2$. Let γ_i $(i = 1, 2)$ be the restriction of γ to K_i. Then

$$\sup [U_\alpha^{\gamma_i}(x) \mid x \in \text{supp } \gamma_i] \leqslant \sup [U_\alpha^{\gamma}(x) \mid x \in \text{supp } \gamma] \leqslant 1$$

so that $\gamma_i \in \mathscr{S}_{K_i}$. Consequently $\gamma(K_i) = \gamma_i(K_i) \leqslant \text{cap } K_i$ and so

$$\text{cap } (K_1 \cup K_2) = \gamma(K_1 \cup K_2) \leqslant \gamma(K_1)+\gamma(K_2) \leqslant \text{cap } K_1+\text{cap } K_2.$$

When $\alpha \leqslant 2$ something more is true. In that case cap is strongly sub-additive, that is,

$$\text{cap}(K_1 \cup K_2)+\text{cap}(K_1 \cap K_2) \leqslant \text{cap } K_1+\text{cap } K_2.$$

The proof that is so is much more difficult than that for Lemma 3.32 and involves a rather delicate and ingenious application of the domination principle. It may be found in Landkov, p. 178.

Finally, cap is 'continuous from the right', that is

LEMMA 3.33. *Given a compact K and given $\varepsilon > 0$, there is an open set $G \supset K$ such that if $K \subset K' \subset G$, K' compact, then*

$$\textit{cap } K' - \textit{cap } K < \varepsilon.$$

If this property did not hold, then there would be a $k > 0$ and a decreasing sequence $\{K_n\}$, $K_n \supset K$, of compacts such that $K = \bigcap_{n=1}^{\infty} K_n$ and

$$\text{cap } K_n \geqslant \text{cap } K+k.$$

Let γ_n be the equilibrium measure of K_n so that

$$\| \gamma_n \|^2 = \gamma_n(\mathbb{R}^p) = \text{cap } K_n.$$

Inspection of the proof of Lemma 3.15 will show that, in the proof of vague compactness there, it is in each case the fact that the sequence $\{\mu_n(f_m)\}$ is bounded in n that is required. This holds for the $\{\gamma_n\}$ and we may therefore conclude that it has a vaguely convergent subsequence $\{\gamma_{r_n}\}$ of which the limit will clearly be a measure μ supported on K (since supp $\mu \subset K_n$ for every n). Also, by (3.16),

$$\| \mu \|^2 \leqslant \lim_{n\to\infty} \| \gamma_{r_n} \|^2 = \lim_{n\to\infty} \| \gamma_n \|^2 = \lim_{n\to\infty} \text{cap } K_n.$$

Also

$$\mu(\mathbb{R}^p) = \lim_{n\to\infty} \gamma_{r_n}(\mathbb{R}^p) \geqslant \text{cap } K+k.$$

Let $m = \mu(\mathbb{R}^p)$. Then $\mu/m \in \mathscr{M}_K$ and so $\| \mu/m \|^2 \geqslant \| \lambda \|^2 = (\text{cap } K)^{-1}$. Thus

$$\| \mu \|^2 \geqslant m^2/\text{cap } K \geqslant (\text{cap } K+k)^2/\text{cap } K \geqslant \text{cap } K+2k.$$

We have therefore shown that, for every n,

$$\text{cap } K_n \geqslant \text{cap } K+k \text{ implies cap } K_n \geqslant \text{cap } K+2k.$$

Repeating this argument i times we find that cap $K_n \geqslant$ cap $K+2^i k$; and, letting $i \to \infty$, we find that cap $K_n = \infty$ for every n. Since this cannot be so, the lemma now follows.

Given any set $E \subset \mathbb{R}^p$, we define its **interior capacity** $\text{cap}_* E$ by

$$\text{cap}_* E = \sup \{\text{cap } K \mid K \subset E \text{ and } K \text{ compact}\}.$$

In particular, this defines the interior capacity of any open set G.

We define the **exterior capacity** $\text{cap}^* E$ of E by

$$\text{cap}^* E = \inf \{\text{cap}_* G \mid G \supset E \text{ and } G \text{ open}\}.$$

It follows immediately that

$$\text{cap}_* E \leqslant \text{cap}^* E. \tag{3.21}$$

For suppose $K \subset E \subset G$, K compact and G open. Then cap $K \leqslant$ $\text{cap}_* G$ and so

$$\sup_{K \subset E} \text{cap } K \leqslant \text{cap}^* G$$

for all $G \supset E$. This now gives (3.21).

LEMMA 3.34.

(i) *$cap_*(a+E) = cap^*E$;*

(ii) *$cap_*(bE) = b^{p-\alpha}\, cap_*E$;*

(iii) *cap_* is increasing.*

The above properties also hold for cap.*

All these follow easily from the definitions.

Finally, a set E is said to be **capacitable,** if

$$\text{cap}_* E = \text{cap}^* E;$$

and, when this is the case, we define the capacity of E, cap E, to be the common value of the interior and exterior capacities.

It is obviously of great interest to know which sets are capacitable. From subsequent considerations it will emerge that, in fact, any Borel set is capacitable. For the moment, however, we will look at the capacitability of open and closed sets. We have first

LEMMA 3.35. *Any open set is capacitable.*

By definition $\text{cap}^* G = \inf \{\text{cap}_* H \mid H \supset G, H \text{ open}\}$. Since $G \supset G$, we have $\text{cap}^* G \leqslant \text{cap}_* G$; which, with (3.21), gives equality.

LEMMA 3.36. *Any compact set is capacitable.*

Somewhat surprisingly, this does not follow directly from the definition. We need to invoke 'continuity from the right'.

Given K and given $\varepsilon > 0$, we can find an open G such that if $K \subset K' \subset G$, then cap K' − cap $K < \varepsilon$; from which it follows that cap G− cap $K < \varepsilon$. But cap $G \geqslant$ cap* K, so cap* $K <$ cap $K+\varepsilon$ for arbitrary ε, and hence cap*$K \leqslant$ cap K. Also, clearly, cap K = cap$_*K$, and hence cap$_*K$ = cap*K so that K is capacitable.

Again, invoking 'continuity on the right', we have

LEMMA 3.37. *Let $\{K_n\}$ be a decreasing sequence of compact sets. Then*

$$\inf_n \operatorname{cap} K_n = \operatorname{cap} \left(\bigcap_{n=1}^{\infty} K_n\right).$$

Let $K = \bigcap_{n=1}^{\infty} K_n$. Then $K_n \supset K$, and so $\inf_n \operatorname{cap} K_n \geqslant \operatorname{cap} K$. Suppose, if possible, that inf cap K_n = cap $K+k$, $k > 0$. Given $G \supset K$, we can find $\delta > 0$ such that the compact $\{x \mid \operatorname{dist}(x, K) \leqslant \delta\} \subset G$ and then N such that, for $n \geqslant N$, K_n is contained in this compact, Hence, for every $G \supset K$ there is K' such that $K \subset K' \subset G$ and cap $K' \geqslant$ cap $K+k$; which contradicts 'continuity from the right'.

Finally, we have an analogous result (Lemma 3.40) about increasing sequences of sets. We shall content ourselves with giving a proof for $\alpha \leqslant 2$; in this we appeal to the fact that then cap is strongly sub-additive. The result still holds for $\alpha > 2$ but then a quite different proof (for which reference may be made to Landkov, p. 193) is required. To prove Lemma 3.40 we need two preliminary results. First, there is

LEMMA 3.38. *Let H_1, H_2 be open. Then, for $\alpha \leqslant 2$,*

$$cap(H_1 \cup H_2)+cap(H_1 \cap H_2) \leqslant cap\, H_1+cap\, H_2.$$

First, the result is immediate if $H_1 \subset H_2$ or $H_2 \subset H_1$. Otherwise, we proceed as follows. Let $\varepsilon > 0$ be given. Then we can find compacts K and L such that $L \subset H_1 \cap H_2, K \subset H_1 \cup H_2, L \subset K$ and

$$\operatorname{cap} L > \operatorname{cap}(H_1 \cap H_2)-\tfrac{1}{2}\varepsilon;\ \operatorname{cap} K > \operatorname{cap}(H_1 \cup H_2)-\tfrac{1}{2}\varepsilon \quad (3.22)$$

(for if $L \not\subset K$ then $K \cup L$ can be taken in place of K).

Choose open sets N_1, N_2 such that $K \cap \partial H_1 \subset N_1 \subset H_2$, $K \cap \partial H_2 \subset N_2 \subset H_1$ and $L \cap N_1 = L \cap N_2 = N_1 \cap N_2 = \phi$. Let $F_i = K \cap \bar{H}_i$, $i = 1, 2$. Then $F_1 \cap N_1 \subset F_2 \setminus N_2$, $F_2 \cap N_2 \subset F_1 \setminus N_1$ so that

$$(F_1 \setminus N_1) \cup (F_2 \setminus N_2) = (F_1 \setminus F_1 \cap N_1) \cup (F_2 \setminus N_2) = F_1 \cup (F_2 \setminus N_2)$$
$$= F_1 \cup (F_2 \setminus F_2 \cap N_2) = F_1 \cup F_2 = K.$$

Thus $K_i = F_i \setminus N_i$, $i = 1, 2$ are compacts such that $K_i \subset H_i$ and $K_1 \cup K_2 = K$ so that

$$\text{cap}(K_1 \cup K_2) + \text{cap}(K_1 \cap K_2) \leqslant \text{cap } K_1 + \text{cap } K_2 \leqslant \text{cap } H_1 + \text{cap } H_2.$$

Also $L \subset F_i$ and $L \cap N_i = \phi$, $i = 1, 2$ so that $L \subset K_1 \cap K_2$. Then, by (3.22) and the last inequality

$$\text{cap}(H_1 \cup H_2) + \text{cap}(H_1 \cap H_2) - \varepsilon < \text{cap } K + \text{cap } L \leqslant \text{cap} H_1 + \text{cap } H_2.$$

Since this holds for every $\varepsilon > 0$ the lemma now follows.

LEMMA 3.39. *Let $B_1 \subset B_2$ and let H_1, H_2 be open with $H_i \supset B_i$, $i = 1, 2$ and such that cap $H_i \leqslant$ cap $B_i + \delta_i$, $i = 1, 2$. Then, for $\alpha \leqslant 2$,*

$$\textit{cap}(H_1 \cup H_2) \leqslant \textit{cap } B_2 + \delta_1 + \delta_2.$$

We have

$$\text{cap}(H_1 \cup H_2) + \text{cap}(H_1 \cap H_2) \leqslant \text{cap } H_1 + \text{cap } H_2 \leqslant \text{cap } B_1 + \text{cap } B_2 + \delta_1 + \delta_2$$

Now $H_1 \cap H_2 \supset B_1$ so cap $B_1 \leqslant \text{cap}(H_1 \cap H_2)$ and hence the lemma follows.

LEMMA 3.40. *Let $\{A_n\}$ be an increasing sequence of sets. Then*

$$\textit{cap}^*\Big(\bigcup_{n=1}^{\infty} A_n\Big) = \textit{sup cap}^* A_n.$$

Let it be stated again that the proof to follow applies only to $\alpha \leqslant 2$.

Suppose first that $\{G_n\}$ is an increasing sequence of open sets, and let $G = \bigcup_{n=1}^{\infty} G_n$. Then $\{\text{cap}^* G_n\}$ is increasing, $G_n \subset G$, and so

$$\lim \text{cap}^* G_n \leqslant \text{cap}^* G. \tag{3.23}$$

Given $\varepsilon > 0$, let $K \subset G$ be a compact such that cap $K > \text{cap}^* G - \varepsilon$. ($\text{cap}_* G = \text{cap}^* G$.) Then $\{G_n\}$ is an open covering of K, and so, by the Borel covering theorem, there is an N such that for $n > N$, $G_n \supset K$ and, consequently,

$$\text{cap}^* G_n \geqslant \text{cap } K > \text{cap}^* G - \varepsilon.$$

Thus $\lim \text{cap}^* G_n \geqslant \text{cap}^* G$, and this, with (3.23), gives

$$\lim_{n \to \infty} \text{cap}^* G_n = \text{cap}^* G. \tag{3.24}$$

Let $J_n \supset A_n$ be open and such that cap $J_n < \text{cap}^* A_n + \varepsilon 2^{-n-1}$. Then

$$\text{cap}(J_1 \cup \ldots \cup J_n) < \text{cap}^* A_n + \varepsilon(1 - 2^{-n}).$$

For suppose the result to be true for n. Let $B_1 = A_n$, $B_2 = A_{n+1}$, $H_1 = J_1 \cup \ldots \cup J_n$ and $H_2 = J_{n+1}$. Then, by Lemma 3.39,

$$\operatorname{cap}(J_1 \cup \ldots \cup J_{n+1}) < \operatorname{cap}^* A_{n+1} + \varepsilon(1 - 2^{-n} + 2^{-n-1})$$

giving the hereditary property. Since the result also holds for $n = 1$ it holds universally.

So, setting $G_n = J_1 \cup \ldots \cup J_n$ we have $\{G_n\}$ increasing and open and $\operatorname{cap} G_n < \operatorname{cap}^* A_n + \varepsilon$ so that, letting n tend to ∞ and noting that ε is arbitrary we have

$$\lim_{n\to\infty} \operatorname{cap} G_n \leqslant \lim_{n\to\infty} \operatorname{cap}^* A_n.$$

By (3.24) the left hand side of the inequality above is equal to $\operatorname{cap}(\bigcup_{n=1}^{\infty} G_n)$ and $\bigcup_{n=1}^{\infty} G_n \supset \bigcup_{n=1}^{\infty} A_n$ so that

$$\operatorname{cap}^*(\bigcup_{n=1}^{\infty} A_n) \leqslant \lim \operatorname{cap}^* A_n$$

Since also $A_n \subset \bigcup_{n=1}^{\infty} A_n$ we now have the reversed inequality and the lemma now follows.

This proven, we can now improve Lemma 3.29 to

LEMMA 3.41. *Any closed set is capacitable.*

Let F be closed. Then $F_n = F \cap \bar{B}_n(0)$ is compact, $\{F_n\}$ is an increasing sequence and $\bigcup_{n=1}^{\infty} F_n = F$ so, by the last lemma and Lemma 3.36

$$\lim_{n\to\infty} \operatorname{cap} F_n = \operatorname{cap}^* F.$$

Since F_n is a compact contained in F, $\operatorname{cap}_* F \geqslant F_n$, and so

$$\operatorname{cap}^* F = \lim_{n\to\infty} \operatorname{cap} F_n \leqslant \operatorname{cap}_* F$$

which, together with (3.21), gives the lemma.

§ 3.7. Generalised Capacity

In 1947 Henri Cartan was forced to leave open the question of whether every Borel set is capacitable for Newtonian potential. Choquet took the matter up in 1953 and, in fact, solved the problem in a much more general setting. We give an account of a small part of his great paper.

Suppose that X is a locally compact Hausdorff space and that ψ is a real-valued set function defined on the compacts in X. Suppose that

(i) ψ is increasing, that is, if $K_1 \subset K_2$ then $\psi(K_1) \leqslant \psi(K_2)$

(ii) ψ is strongly sub-additive, that is

$$\psi(K_1 \cup K_2)+\psi(K_1 \cap K_2) \leqslant \psi(K_1)+\psi(K_2)$$

(iii) ψ is continuous from the right, that is, given K and given $\varepsilon > 0$, there is an open set G such that if $K \subset K' \subset G$, K' compact then $\psi(K')-\psi(K) < \varepsilon$.

Then ψ is said to be a **Choquet capacity.** Define the interior capacity $\psi_*(A)$ of a set A by

$$\psi_*(A) = \sup_{K \subset A} \psi(K)$$

and then the exterior capacity $\psi^*(A)$ by

$$\psi^*(A) = \inf_{G \supset A} \psi_*(G)$$

and say that a set A is ψ-capacitable if $\psi_*(A) = \psi^*(A)$. It is then clear that we can use the arguments of §3.6 to show that any open set and any closed set are ψ-capacitable. Furthermore we can show that ψ satisfies Lemma 3.37 and ψ^* satisfies Lemma 3.40. It is this that leads us to a further generalisation. Suppose now that ϕ is a set function defined on all the subsets of X and taking numerical values which has the following properties:

(*a*) *If* $A_1 \subset A_2$ *then* $\phi(A_1) \leqslant \phi(A_2)$.

(*b*) *If* $\{A_n\}$ *is an increasing sequence of sets,* $\phi(\cup A_n) = \sup_n \phi(A_n)$.

(*c*) *If* $\{K_n\}$ *is a decreasing sequence of compact sets then* $\phi(\cap K_n) = \inf_n \phi(K_n)$.

Then ϕ is said to be a **generalised capacity.** Furthermore, we say that a set A is ϕ-capacitable if

$$\phi(A) = \sup_{K \subset A} \phi(K), \ K \text{ compact}$$

We note that cap* is a generalised capacity and that the criterion of ϕ-capacitability here coincides with that of § 3.6 since

$$\sup_{K \subset A} \text{cap}^* K = \sup_{K \subset A} \text{cap } K = \text{cap}_* A.$$

and that there are similar considerations for ψ^*.

Suppose that to every finite set $\{n_1, \ldots, n_p\}$ of positive integers there corresponds a compact set $K_{n_1, \ldots, n_p}$. Let

$$A = \{\cup K_{n_1} \cap K_{n_1, n_2} \cap \ldots \mid n_i = 1, 2, 3, \ldots, i = 1, 2, 3, \ldots\}. \quad (3.25)$$

The systems of compacts $\{K_{n_1, n_2, \ldots}\}$ is said to generate A, and any set A generated by such a system of compacts is said to be **analytic.**

LEMMA 3.42. *Suppose that X has a countable base of neighbourhoods. Then the family $\mathscr{A}$ of all analytic subsets of X is such that*

(*a*) *Every closed set belongs to $\mathscr{A}$.*

(*b*) If $\{A_k\}$ *is a sequence of sets with* $A_k \in \mathscr{A}$, *then* $\bigcap_{k=1}^{\infty} A_k \in \mathscr{A}$ *and* $\bigcup_{k=1}^{\infty} A_k \in \mathscr{A}$.

Since X has a countable base of neighbourhoods and is locally compact, $X = \bigcup_{n=1}^{\infty} X_n$ where X_n is compact. Then, if F is closed $F = \bigcup_{n=1}^{\infty} K_n$ with $K_n = F \cap X_n$, so that F is the union of a countable family of compacts, and so $F \in \mathscr{A}$. This gives (*a*).

Let $B_k = \overline{N}_k$ where $\{N_k\}$ is an increasing neighbourhood base of relatively compact open sets for E and let

$$B_{k, n_1, \ldots, n_p} = K^k_{n_1, n_2, \ldots, n_p}.$$

Then the system $\{B_{k, n_1, \ldots, n_p}\}$ generates $\bigcup_{k=1}^{\infty} A_k$ and so this latter set is analytic.

Next

$$\bigcap_{k=1}^{\infty} A_k = \cup \{K_{n_1} \cap K_{n_1, n_2} \cap \ldots \mid n_2 = 1, 2, 3, \ldots; i = 1, 2, 3, \ldots\}$$

where

$$K_{n_1, n_2, \ldots, n_p} = \bigcap_{k=1}^{\infty} K^k_{n_1, n_2, \ldots, n_p}$$

and so the system $\{K_{n_1, n_2, \ldots, n_p}\}$ of compacts generates $\bigcap_{k=1}^{\infty} A_k$ which is therefore analytic. This gives (*b*).

We now have a theorem which shows that the set of all ϕ-capacitable sets is very large.

THEOREM 3.43. *Let X have a countable base of neighbourhoods. Then every analytic set is ϕ-capacitable.*

Let A be given by (3.25). Define S_1^m by

$$S_1^m = \{\cup K_{n_1} \cap K_{n_1, n_2} \cap \ldots \mid n_1 \leqslant m;\ n_i = 1, 2, \ldots, i = 2, 3, \ldots\}.$$

Then $\{S_1^m\}$ is an increasing sequence with union A and so, given $\varepsilon > 0$, we can, by property (*b*) for a generalised capacity, choose m_1 so that

$$\phi(S_1^{m_1}) \geqslant \phi(A) - \tfrac{1}{2}\varepsilon.$$

Let S_2^m be given by

$$S_2^m = \{\cup K_{n_1} \cap K_{n_1, n_2} \cap \ldots \mid n_1 \leqslant m_1;\, n_2 \leqslant m_2;$$
$$n_i = 1, 2, \ldots, i = 3, 4, \ldots\}.$$

Then $\{S_2^m\}$ is an increasing sequence with union $S_1^{m_1}$, and so we can choose m_2 so that

$$\phi(S_2^{m_2}) \geqslant \phi(S_1^{m_1}) - \tfrac{1}{4}\varepsilon.$$

Suppose that, in general, we have already found $m_1, \ldots, m_{p-1}$ and $S_1^{m_1}, \ldots, S_{p-1}^{m_{p-1}}$ such that, setting $S_0^{m_0} = A$,

$$\phi(S_r^{m_r}) \geqslant \phi(S_{r-1}^{m_{r-1}}) - \varepsilon\, 2^{-r},\ r = 1, \ldots, p-1.$$

Then set

$$S_p^m = \{\cup K_{n_1} \cap K_{n_1, n_2} \cap \ldots \mid n_i \leqslant m_i,\, i = 1, \ldots, p-1;$$
$$n_i = 1, 2, \ldots, i = p, p+1, \ldots\}.$$

Then $\{S_p^m\}$ increases to $S_{p-1}^{m_{p-1}}$, and so we can find m_p such that

$$\phi(S_p^{m_p}) \geqslant \phi(S_{p-1}^{m_{p-1}}) - \varepsilon\, 2^{-p}.$$

Thus the induction step. So we can construct, for every p, $S_p^{m_p}$ such that

$$\phi(S_p^{m_p}) - \phi(A) = \sum_{r=1}^{p} (\phi(S_r^{m_r}) - \phi(S_{r-1}^{m_{r-1}})) \geqslant -\varepsilon \sum_{r=1}^{\infty} 2^{-r} = -\varepsilon$$

and hence, for every p,

$$\phi(S_p^{m_p}) \geqslant \phi(A) - \varepsilon.$$

Let

$$F_p = \{\cup K_{n_1} \cap K_{n_1, n_2} \cap \ldots \cap K_{n_1, n_2, \ldots, n_p} \mid n_i \leqslant m_i,\, i = 1, \ldots, p\}$$

and let $F = \bigcap_{p=1}^{\infty} F_p$. Then F_p is compact and contains $S_p^{m_p}$, so that, for every p,

$$\phi(F_p) \geqslant \phi(A) - \varepsilon.$$

By property (c) for a generalised capacity, we have

$$\phi(F) = \lim_{p \to \infty} \phi(F_p) \geqslant \phi(A) - \varepsilon.$$

Also, $F \subset A$. For suppose $x \in F$. Then, to every p there correspond integers $n_{ip} \leqslant m_i$ such that

$$x \in K_{n_{1p}} \cap K_{n_{1p}, n_{2p}} \cap \ldots \cap K_{n_{1p}, \ldots, n_{pp}}.$$

We may choose $\{p_k\}$ such that $t_i = \lim_{k \to \infty} n_{ip_k}$ exists for all i, since $n_{ip} \leqslant m_i$. Then

$$x \in K_{t_1} \cap K_{t_1, t_2} \cap \ldots$$

Consequently $x \in A$, and thus $F \subset A$.

Thus F is compact, $F \subset A$ and $\phi(F) \geqslant \phi(A) - \varepsilon$. Therefore

$$\phi(A) = \sup\{\phi(K) \mid K \subset A \text{ and } K \text{ compact}\},$$

so that A is ϕ-capacitable and Theorem 3.43 is demonstrated.

It is clear that the complement of an analytic set is analytic, and so $\mathscr{A}$ is a tribe containing the closed sets. Since the Borel tribe is the smallest such tribe, it follows that any Borel set is analytic.

Thus, in particular, any Borel set in $\mathbb{R}^p$ is α-capacitable. Also, any $K_{\sigma\delta}$ set, i.e. any set which is the intersection of a countable family of sets each of which is the union of a countable family of compact sets, is clearly analytic and so ϕ-capacitable for any generalised capacity.

Chapter 4

THE DIRICHLET PROBLEM

§ 4.1. Abstract Harmonic Functions

One of the most important and most beautiful results in Potential Theory is the generalised solution of the Dirichlet Problem. As we saw in Chapter 2, in the course of the discussion on the Poisson Integral, the problem is the following:

Given an open set $G \subset \mathbb{R}^p$ and a function f defined and continuous in ∂G, to find a function u harmonic in G such that, for all $\xi \in \partial G$,

$$\lim_{x \to \xi} u(x) = f(\xi). \tag{4.1}$$

The problem has a long and interesting history. In 1828 Green offered a proof based on physical intuition. In 1840 Gauss offered a somewhat more mathematical solution, as did, in 1847, Lord Kelvin, initiating what became to be known as Dirichlet's Principle. The reader is referred to a very fine historical introduction in Kellogg, in which he gives an argument which shows that the Dirichlet Problem can be solved in every case in which the Dirichlet Integral has a minimum under the given conditions. This leaves open the large question of whether the Dirichlet Integral always has a minimum, and in 1870 Weierstrass showed that this need not be the case. It was not until 1899 that Hilbert showed, under proper conditions on the region, boundary values and the functions admitted that the principle could be proved reliable.

The matter was taken further by Neumann and Poincaré. The former needed to make the assumption that the region was convex; while the latter was able to produce an existence theorem, provided the boundary satisfied an exterior sphere condition. At this stage, to quote Kellogg, 'So far it was generally believed that the Dirichlet Problem was solvable for any region, and that the limitations of generality were inherent in the methods rather than the problem itself. It was Zaremba (in 1911) who first pointed out that there were regions for which the problem was not possible.'

A quite different approach to the problem was initiated by Perron in 1923, speedily added to by Wiener in 1925; and, after further additions and improvement by many others, the whole was brought to a definitive solution of great power and beauty by Brelot in 1939. Now, any restrictions on the boundary ∂G are abandoned, and it is shown that for any bounded open set G, and for any continuous function f in ∂G, one can construct a harmonic function H_f^G which will be the solution to the Dirichlet Problem if there is one. Furthermore, although (1) will, in general, no longer hold for *all* $\xi \in \partial G$, it will hold for 'almost' all $\xi \in \partial G$. And all this is about a boundary value problem for a particular partial differential equation, the Laplace Equation. The question arises whether similar results might be expected for other partial differential equations, notably elliptic partial differential equations.

It will be recalled that in Chapter 1 we began working with a ground space considerably more general than $\mathbb{R}^p$. Indeed, we passed from a locally compact Hausdorff space to $\mathbb{R}^p$ only as soon as it became necessary to talk about differentiation. But even if, perhaps, one did not wish to discuss the Laplacian, and so could possibly sidestep the necessity to talk about differentiation, one would still need to talk of spherical and ball measures and so would need a metric and a translation invariant measure. But a locally compact space which has both a metric and linear structure must effectively be $\mathbb{R}^p$.

So one could pose the question: 'How can one talk of harmonic functions defined on a locally compact space which, of course, does not have either algebraic or metric structure?' The answer, due to Doob and Tautz in the first instance and later to Brelot, is a surprising one, partly because, at first sight, it seems clumsy and somewhat contrived. But it is, in fact, very efficient and, like all good generalisations, produces deep results in a number of disparate fields. To quote Bauer, 'der Laplaceschen Differentialgleichung stets als Leitlinie vor Augen haben möge' is necessary to understand why we are led to lay down the axioms we do in what follows.

The Laplace equation of course gives rise to harmonic functions, and we should perhaps first remind ourselves of some of the properties of these.

First, if u and v are harmonic in an open set $G \subset \mathbb{R}^p$ then, for $\alpha, \beta \in \mathbb{R}$, so is $\alpha u + \beta v$. Thus **the set of functions harmonic in G form a real linear space.**

Second, **if u is harmonic in G, and G' is an open set contained in G, then u is harmonic in G'. Furthermore, if u is harmonic in a neighbourhood of every point of G, then it is harmonic in G.**

These are the properties that will be abstracted when we set up Axiom 1 below.

Next, the Poisson Integral solves the Dirichlet Problem for any open ball and **the family of all open balls in $\mathbb{R}^p$ constitute a basis for the topo-**

logy in $\mathbb{R}^p$, i.e. every open set in $\mathbb{R}^p$ is a union of open balls. This notion of a plentiful supply of 'nice' open sets from a Dirichlet point of view prompts the definition of 'regular set' below and the formulation of Axiom 2.

Finally, we have, as we have seen, that remarkable result (Theorem 2.24) that **an up-directed family of functions each harmonic in an open domain G either has its supremum harmonic in G or identically $+\infty$ in G.** It is this which inspires Axiom 3 below.

These three axioms will enable us to set up in business, as it were, and to construct a quite rich structure. Later, to obtain precise results it will be necessary to introduce an AXIOM P, an AXIOM D and make the assumption that the space X has a countable base of neighbourhoods. These later axioms will be introduced when they become necessary.

We turn now to the statement of Axioms 1, 2 and 3. Suppose that X is a locally compact Hausdorff space which is, in addition, connected. This requirement that X also be connected is a restriction which is more apparent than real since, if X were not connected, each component would go its separate way in that there would be a harmonic structure in each component.

AXIOM 1. *Given any open set $G \subset X$ there is a real linear subspace $\mathscr{H}_G$ of $\mathscr{C}(G)$ such that*

(1) *If $G' \subset G$ is open and $f \in \mathscr{H}_G$ then the restriction of f to G',* $\operatorname{rest}_{G'} f \in \mathscr{H}_{G'}$.

(2) *If $f \in \mathscr{C}(G)$ and if, for each $x \in G$, there is a neighbourhood N of x such that* $\operatorname{rest}_N f \in \mathscr{H}_N$, *then $f \in \mathscr{H}_G$.*

Axiom 1 can be described shortly, if technically, by saying that the mapping $G \to \mathscr{H}_G$ is a pre-sheaf of real continuous functions on X. We shall refer to $\mathscr{H}_G$ as **the set of functions harmonic in G.**

A connected open set $B \subset X$ is said to be a **regular set** if it is relatively compact, $\partial B \neq \phi$, and, given $f \in \mathscr{C}(\partial B)$, there is a unique member of $\mathscr{H}_B$, denoted by H_f^B, such that

$$\lim_{x \in B,\, x \to \xi} H_f^B(x) = f(\xi) \text{ for all } \xi \in \partial B$$

and if, furthermore, when $f(\xi) \geqslant 0$, $\xi \in \partial B$, then $H_f^B(x) \geqslant 0$, $x \in B$.

Then

AXIOM 2. *The family of regular sets in X is a basis for the topology of X.*

AXIOM 3. *Let $\{u_\alpha\}_{\alpha \in A}$ be an up-directed family of functions with $u_\alpha \in \mathscr{H}_G$, where G is a connected open set. Then either $\sup_A u_\alpha$ is harmonic in G, i.e. $\in \mathscr{H}_G$, or $\sup_A u_\alpha = +\infty$ everywhere in G.*

It is an immediate consequence of Axiom 3 and Axiom 1 that, if $\{u_\alpha\}_{\alpha \in A}$ is a down-directed family of functions each of which is harmonic in G, then either $\inf_A u_\alpha$ is harmonic in G, or $\inf_A u_\alpha = -\infty$ everywhere in G.

We are now in a position to see to what extent this theory furnishes a minimum principle for harmonic functions.

THEOREM 4.1. *Suppose that u is harmonic and non-negative in a domain G. Then either $u(x) > 0$ for all $x \in G$, or $u(x) = 0$ for all $x \in G$.*

Consider the sequence $\{nu\}_{n=1,2,\ldots}$. Since it is an increasing sequence, we have, from Axiom 3, that either $\lim_{n\to\infty} nu(x) = +\infty$ for all $x \in G$ or $\lim_{n\to\infty} nu(x)$ is harmonic in G. Thus, if for some $a \in G$, $u(a) = 0$ then $\lim_{n\to\infty} nu(a) = 0$, and so the first possibility cannot hold. Hence $\lim_{n\to\infty} nu(x)$ is finite for $x \in G$, and this requires $u(x) = 0$ for all $x \in G$.

It is almost instinctive to make the next step and to assert that a harmonic function attains its minimum in a domain only if it is constant there. But this involves the assumption that a constant is harmonic, and it is an unwelcome quirk of the general theory that this need not be true. Indeed, as Hervé has shown, the solution of the uniformly elliptic differential equation

$$\sum_{i,j=1}^{p} a_{ij}(x) \frac{\partial^2 u}{\partial x_i \, \partial x_j} + \sum_{i=1}^{p} b_i(x) \frac{\partial u}{\partial x_i} + c(x)u = 0,$$

for which the coefficients a_{ij}, b_i and c are locally Lipschitz in a domain G, defines a system satisfying Axioms 1, 2, 3 (and, for that matter, Axiom P and Axiom D). Clearly if $c(x)$ is not identically zero, then the constants are not harmonic in this sense.

If, however, there is a strictly positive harmonic function h in X, that is $h \in \mathscr{H}_X$, we can sometimes, with profit, consider the family $\{f/h \mid f \in \mathscr{H}_X\}$. Then this new family of so-called h-harmonic functions satisfies Axioms 1, 2 and 3 and contains all the constants, since $h \in \mathscr{H}$ and $h/h = 1$. Consequently the following result is of interest.

LEMMA 4.1.1. *Given a regular set B, there is a function $h \in \mathscr{H}_B$ such that $h(x) > k > 0$ for all $x \in B$.*

Now $H_1^B(x) \to 1$ as $x \to \xi \in \partial B$, and $H_1^B(x) \geqslant 0$. Also, $H_1^B(x) > 0$, since, if it were not so, we would have $H_1^B(x) = 0$ in B, by Theorem 4.1, and this would contradict $H_1^B(x) \to 1$ as $x \to \xi$. Hence $H_1^B(x) > 0$ in $\bar{B}$, and so, since $\bar{B}$ is compact,

$$\inf_{\bar{B}} H_1^B(x) > 0,$$

which gives the result.

We have seen that, in dealing with the Laplace-harmonic functions, we make considerable use of the notion of spherical measure. For a function f which is Laplace-harmonic in G, and for every ball $B_R(a)$ such that $\bar{B}_R(a) \subset G$, we have

$$f(a) = \int f(y)\, \sigma_R^a(dy).$$

Although the regular set could conceivably play the role that the ball plays in $\mathbb{R}^p$, there is no way of singling out a point $a \in B$ which would be its 'centre'. Fortunately, the Poisson Integral furnishes a generalisation of the mean property in the following way:

Let the measure σ_x^B be defined by

$$\sigma_x^B(E) = \int_{E \cap S_R(a)} K_R(x, t)\, \sigma_R^a(dt).$$

Then

$$(I_R^a f)(x) = \int f(y)\, \sigma_x^B(dy)$$

whenever $x \in B_R(a)$.

It is this last relation which suggests the following formulation in the general theory. Let B be any regular set in X. Let $f, g \in \mathscr{C}(\partial B)$. Then $\alpha f + \beta g \in \mathscr{C}(\partial B)$. Now, for $\xi \in \partial B$,

$$\lim_{x \to \xi} H_f^B(x) = f(\xi);\ \lim_{x \to \xi} H_g^B(x) = g(\xi) \text{ and } \lim_{x \to \xi} H_{\alpha f + \beta g}^B = (\alpha f + \beta g)(\xi).$$

Also $\alpha H_f^B + \beta H_g^B \in \mathscr{H}_B$ and

$$\lim_{x \to \xi} (\alpha H_f^B + \beta H_g^B)(x) = (\alpha f + \beta g)(\xi) \tag{4.2}$$

Since there is just one element of $\mathscr{H}_B$ satisfying (4.2) we have

$$H_{\alpha f + \beta g}^B = \alpha H_f^B + \beta H_g^B.$$

Thus, for each $x \in B, f \to H_f^B(x)$ is a positive linear functional on $\mathscr{C}(\partial B)$ and so, by the Riesz Representation Theorem (1.20) there is a unique measure σ_x^B such that

$$H_f^B(x) = \int f(y)\, \sigma_x^B(dy)$$

for every $f \in \mathscr{C}(\partial B)$ and every $x \in B$.

The measure σ_x^B is known as **the harmonic measure for B at x.**

We have seen that, in the classical case (Lemma 2.16) $H_1^B(x) = 1$ for $x \in B$. This is only true in the general case when constants are harmonic; for then, since $1 \in \mathscr{H}_B$, and H_1^B is unique, we have $H_1^B(x) = 1$. When this is not the case, Lemma 2.16 must be replaced by

LEMMA 4.2. *Given $0 < \varepsilon < 1$ and $a \in X$, there is a neighbourhood $N(\varepsilon)$ of a such that, when $a \in B \subset N(\varepsilon)$,*

$$|\, H_1^B(a) - 1 \,| < \varepsilon.$$

Let B_0 be a regular set containing a, and let $h \in \mathscr{H}_{B_0}$ be such that $\inf_{B_0} h > 0$ and $h(a) = 1$. Then, given $B \subset B_0$, and setting $f = \operatorname{rest}_{\partial B} h$, we have

$$h(a) = H_f^B(a) = \int h(\xi)\, \sigma_a^B(d\xi)$$

since $\lim_{x \in B,\, x \to \xi} h(x) = f(\xi) = h(\xi)$ and H_f^B is unique.

There is a neighbourhood $N(\varepsilon)$ of a such that, for $x \in N$,

$$| h(x) - h(a) | < \tfrac{1}{2}\varepsilon$$

and so

$$| 1 - \int \sigma_a^B(d\xi) | = | \int (h(\xi) - h(a))\, \sigma_a^B(d\xi) | \leqslant \int | h(\xi) - h(a) |\, \sigma_a^B(d\xi)$$

and, for $B \subset N$, this last is less than $\tfrac{1}{2}\varepsilon \int \sigma_a^B(d\xi)$. Consequently, for $B \subset N$,

$$(1 + \tfrac{1}{2}\varepsilon)^{-1} \leqslant \int \sigma_a^B(d\xi) \leqslant (1 - \tfrac{1}{2}\varepsilon)^{-1}$$

from which the lemma follows.

We have already started by postulating regular sets, sets such that any continuous data on the boundary of such a set has associated with it a function harmonic in the set. The introduction of the harmonic measure makes it possible to allow much more general data on the boundary.

THEOREM 4.3. *Given a regular set B and a non-negative function f defined on ∂B*

$$\overline{\int} f(\xi)\, \sigma_x^B(d\xi)$$

is either harmonic in B, or is identically $+\infty$ in B. Furthermore, the integrability of f with respect to σ_x^B is independent of x in B and, for such an integrable f,

$$\int f(\xi)\, \sigma_x^B(d\xi)$$

is harmonic in B.

Let ψ be bounded below and l.s.c. in B. Then, by Lemma 1.17,

$$\int \psi(\xi)\, \sigma_x^B(d\xi) = \sup \{ \int \phi(\xi)\, \sigma_x^B(d\xi) \mid \phi \in \mathscr{C}(\partial B) \text{ and } \phi \leqslant \psi \}.$$

Now $\int \phi(\xi)\, \sigma_x^B(d\xi)$ is harmonic in B, and the family of ϕ is up-directed on B, and so the family of harmonic functions on the right is up-directed. Hence, by Axiom 3, the left-hand side is either harmonic in B or is identically $+\infty$ there.

Now let f be any non-negative function defined in ∂B. Then

$$\overline{\int} f(\xi)\, \sigma_x^B(d\xi) = \inf \{ \int \psi(\xi)\, \sigma_x^B(d\xi) \mid \psi \text{ l.s.c. in } \partial B \text{ and } \psi \geqslant f \}.$$

Then the family of ψ on the right hand side is down-directed, and so the family of integrals on the right consists either entirely of identically $+\infty$ functions or contains a down-directed family of harmonic functions. So, by the down-directed property, either the integral on the left is $+\infty$ identically in B, or is harmonic in B.

A similar argument serves to show that $\underline{\int} f(\xi)\, \sigma_x^B(d\xi)$ is either $+\infty$ in B, or harmonic in B.

Consequently, $(\overline{\int} - \underline{\int})\, f(\xi)\, \sigma_x^B(d\xi)$ is either harmonic and non-negative in B, or is identically $+\infty$ in B, or is undefined everywhere in B. If now f is σ_x^B-integrable for some $x \in B$ then both $\overline{\int}$ and $\underline{\int}$ are finite for that x, and are equal. Also then $\overline{\int} - \underline{\int}$ is harmonic and non-negative in B and attains the value 0 at a point in B. By Theorem 4.1, $\overline{\int} - \underline{\int} = 0$ everywhere in B, and so f is integrable for all $x \in B$. So $\int f(\xi)\, \sigma_x^B(d\xi)$ exists for all $x \in B$ and is harmonic there.

Corollary 4.3.1. *On ∂B, the sets of harmonic measure zero are independent of $x \in B$.*

Let $E \subset \partial B$, and let χ_E be the indicator function for E. Now $\overline{\int} \chi_E(\xi)\, \sigma_x^B(d\xi)$, since it is finite, must be harmonic in B. Since it is non-negative, it will be zero for all $x \in B$ if it is zero for any $x \in B$. Consequently $\sigma_x^B(E) = 0$ for all $x \in B$ or for no $x \in B$.

Corollary 4.3.2. *Let B be a regular set. Then any neighbourhood of any point in ∂B has non-zero σ_x^B-measure.*

Suppose that $\xi \in \partial B$ and that N is a neighbourhood of ξ which has zero σ_x^B-measure. There is, by Urysohn's Lemma, a continuous function f in ∂B taking value 1 at ξ and taking value 0 outside N with $0 \leqslant f(\eta) \leqslant 1$ for $\eta \in \partial B$. Now $\int f(\eta)\, \sigma_x^B(d\eta)$ is zero in B by Corollary 4.3.1, but must tend to 1 at ξ, and this is impossible. The corollary thus follows.

§ 4.2. Hyper- and hypoharmonic Functions

Given an open set $G \subset X$, a numerical-valued function u is said to be **hyperharmonic in G**, if

(1) $u(x) > -\infty$ for all $x \in G$;

(2) u is l.s.c. in G;

(3) given a regular set B such that $\overline{B} \subset G$, and given $x \in B$,

$$u(x) \geqslant \int u(\xi)\, \sigma_x^B(d\xi).$$

We may notice that (3) is not the analogue of the supermean inequality we use in the classical case (and how could it be, since B does not have a centre or radius?) but rather the analogue of Theorem 2.27 relating a superharmonic function to the Poisson Integral of its restriction to the sphere.

We denote the set of functions hyperharmonic in G by $\mathscr{H}_G^*$ and we remark that $\mathscr{H}_G^*$ is, at any rate, non-void since the function which is identically $+\infty$ belongs to $\mathscr{H}_G^*$.

A function u will be said to be **hypoharmonic in G** if $-u \in \mathscr{H}_G^*$; and we shall denote the set of functions hypoharmonic in G by $(-\mathscr{H}_G^*)$.

It now follows, as in the classical case, that

$$\mathscr{H}_G = \mathscr{H}_G^* \cap (-\mathscr{H}_G^*)$$

since, if $u \in \mathscr{H}_G$, then u is continuous in G, and we have equality in (3). Since the right-hand side in (3) is harmonic in B, $u \in \mathscr{H}_B$ for every $\bar{B} \subset G$ and so $u \in \mathscr{H}_G$.

The important properties of hyperharmonic functions detailed in Theorem 2.1 hold also for generalised hyperharmonic functions. Although the argument employed is quite what would be expected as we pass from the classical case to the abstract, we shall give it in detail since it is technically convenient to incorporate an unexpected variation at this point. We shall carry through the proofs assuming, in place of condition (3) for hyperharmonicity, the weaker condition

(3′) for any $a \in G$ there is a base of regular neighbourhoods $\{B_\alpha(a)\}_{\alpha \in A}$ of a such that $\bar{B}_\alpha(a) \subset G$ and

$$u(a) \geqslant \int u(\xi)\, \sigma_a^{B_\alpha(a)}(d\xi).$$

Thus the family $\{B_\alpha(a)\}_{\alpha \in A}$ is such that, firstly, each $B_\alpha(a)$ is a regular set, and secondly, given any open set N containing a there is a $B_\alpha(a) \subset N$.

When u satisfies (1), (2) and (3′), we shall say that u is **locally hyperharmonic** in G and write $u \in \mathscr{L}\mathscr{H}_G^*$. Also, we shall say that u is **locally hypoharmonic in G** if $-u \in \mathscr{L}\mathscr{H}_G^*$, and we shall write $u \in (-\mathscr{L}\mathscr{H}_G^*)$. Clearly, hyperharmonicity implies local hyperharmonicity. We shall show later than the converse implication is also true.

THEOREM 4.4. (1) *Suppose that* $f_r \in \mathscr{L}\mathscr{H}_G^*$ $(\in \mathscr{H}_G^*)$, $r = 1, \ldots, k$ *and that* λ_r, $r = 1, \ldots, k$ *are non-negative constants. Then* $\sum_{r=1}^{k} \lambda_r f_r \in \mathscr{L}\mathscr{H}_G^*$ $(\in \mathscr{H}_G^*)$.

(2) *Suppose that* $f_r \in \mathscr{L}\mathscr{H}_G^*$ $(\in \mathscr{H}_G^*)$, $r = 1, \ldots, k$. *Then* $\inf_{1 \leqslant r \leqslant k} f_r \in \mathscr{L}\mathscr{H}_G^*$ $(\in \mathscr{H}_G^*)$.

(3) *Suppose that* $\{f_\gamma\}_{\gamma \in \Gamma}$ *is an up-directed family and* $f_\gamma \in \mathscr{L}\mathscr{H}_G^*$ $(\in \mathscr{H}_G^*)$. *Then* $\sup f_\gamma \in \mathscr{L}\mathscr{H}_G^*$ $(\in \mathscr{H}_G^*)$.

First, by Theorem 1.1, $g(x) = \sum \lambda_r f_r(x)$ is l.s.c. in G. Next, $g(x) > -\infty$ in G and, finally,

$$g(a) = \sum \lambda_r f_r(a) \geqslant \sum \lambda_r \int f_r(\xi)\, \sigma_a^{B_\alpha(a)}(d\xi) = \int g(\xi)\, \sigma_a^{B_\alpha(a)}(d\xi)$$

so that $g \in \mathscr{L}\mathscr{H}_G^*$. Since the family of all regular B containing a and such that $\bar{B} \subset G$ is also a neighbourhood base for a, the result for hyperharmonic functions also follows. Thus (1).

In fact, to prove (1) we need to be able to assert that there is *one* base of regular neighbourhoods $\{B_\alpha(a)\}_{\alpha \in A}$ such that (3′) holds for every one of f_r, $r = 1, 2, \ldots k$. Let us assume for the moment that this is so and note that the assumption is needed only for (1). It is in fact a temporary expedient since, if $\mathscr{L}\mathscr{H}_G^* = \mathscr{H}_G^*$ then $f_r \in \mathscr{H}_G^*$ if it is locally hyperharmonic. The neighbourhood base for $\mathscr{H}_G^*$ does have the required property and so $g \in \mathscr{H}_G^*$ and hence $g \in \mathscr{L}\mathscr{H}_G^*$ so that, indeed, finally, (1) will hold even without this assumption.

Again, by Theorem 1.1, $g(x) = \inf_{1 \leqslant r \leqslant k} f_r(x)$ is l.s.c. in G and $g(x) > -\infty$ in G. Next, for $a \in G$, choose s so that $f_s(a) = g(a)$. Then

$$g(a) = f_s(a) \geqslant \int f_s(\xi)\, \sigma_a^{B_\alpha(a)}(d\xi) \geqslant \int g(\xi)\, \sigma_a^{B_\alpha(a)}(d\xi)$$

so that $g \in \mathscr{L}\mathscr{H}_G^*$. We get the $\mathscr{H}_G^*$ result as before. Thus (2).

By Theorem 1.6 $g(x) = \sup_\Gamma f_\gamma(x)$ is l.s.c. in G and, clearly, $g(x) > -\infty$ in G. Finally, for $a \in G$ and for $\gamma \in \Gamma$

$$g(a) \geqslant f_\gamma(a) \geqslant \int f_\gamma(\xi)\, \sigma_a^{B_\alpha(a)}(d\xi).$$

By Theorem 1.18

$$\sup_\Gamma \int f_\gamma(\xi)\, \sigma_a^{B_\alpha(a)}(d\xi) = \int g(\xi)\, \sigma_a^{B_\alpha(a)}(d\xi)$$

and hence

$$g(a) \geqslant \int g(\xi)\, \sigma_a^{B_\alpha(a)}(d\xi)$$

so that $g \in \mathscr{L}\mathscr{H}_G^*$. We get the $\mathscr{H}_G^*$ result as before. Thus (3).

For $(-\mathscr{L}\mathscr{H}_G^*)$ and $(-\mathscr{H}_G^*)$ we get similar results. (1) stands unchanged, in (2) 'sup' takes the place of 'inf' and, in (3), we have 'down-directed' instead of 'up-directed'.

One could not expect to have an exact analogue of the 'all-or-none' principle embodied in Theorem 2.2 if only because there is no obvious generalisation for (Lebesgue) measure zero in a general locally compact space and, indeed, there cannot be a translation-invariant measure since there is no linear structure in X to provide a translation. But there is, nevertheless, a modified all-or-none principle.

THEOREM 4.5. *Suppose that G is a domain, that $u \in \mathcal{L}\mathcal{H}_G^*$ $(-\mathcal{L}\mathcal{H}_G^*)$ and that $u(x) = +\infty$ $(-\infty)$ for x in an open subset of G. Then u is $+\infty$ $(-\infty)$ identically in G.*

Let

$$E = \{x \in G \mid u(y) = +\infty \text{ in a neighbourhood of } x\}.$$

Then E is a non-void open subset of G. Suppose that $E \neq G$. Then there will be a non-void component E_1 of E and $\partial E_1 \cap G \neq \phi$ (for otherwise $(E_1 \cap G) \cup (G \setminus E_1)^0 = G$, which would contradict the connectedness of G). Let $x_0 \in \partial E_1 \cap G$ and suppose that $B_\alpha(x_0)$ is a regular set such that $\bar{B}_\alpha(x_0) \subset G$ and

$$u(x_0) \geqslant \int u(\xi)\, \sigma_{x_0}^{B_\alpha(x_0)}(d\xi). \tag{4.3}$$

Also we may suppose that $B_\alpha(x_0)$ does not contain E_1. So let $x_1 \in E_1 \cap \partial B_\alpha$. Then u is $+\infty$ in a neighbourhood of x_1. This neighbourhood has non-zero $\sigma_{x_0}^{B\alpha}$-measure and so the integral in (4.3) is $+\infty$. Since integrability is independent of the point in B_α we have $u(x) = +\infty$ in $B_\alpha(x_0)$. Hence $x_1 \in E$; and this is a contradiction. Thus $E = G$ and the theorem follows.

This theorem leads us to make the following definition.

A function $u \in \mathcal{L}\mathcal{H}_G^*$ $(-\mathcal{L}\mathcal{H}_G^*)$ is said to be locally superharmonic (subharmonic) in G if it takes finite values at at least one point of each of the components of G.

Super- and subharmonicity in G are similarly defined with $\mathcal{L}\mathcal{H}_G^*$ replaced by $\mathcal{H}_G^*$ in the definition above.

THEOREM 4.6. (*The minimum principle for hyperharmonic functions.*)

(1) *Suppose that G is a domain, that $u \in \mathcal{L}\mathcal{H}_G^*$ and that $u \geqslant 0$ in G. Then either $u > 0$ everywhere in G, or $u = 0$ everywhere in G.*

(2) *Suppose that G is a relatively compact open set and that there is $h \in \mathcal{H}_G^*$ such that $\inf_G h > 0$ (which is so if G is regular). Then, if $u \in \mathcal{L}\mathcal{H}_G^*$ and $\liminf_{x \in G, x \to \xi} u(x) \geqslant 0$ for $\xi \in \partial G$ then $u(x) \geqslant 0$ in G.*

Suppose that for some $x_0 \in G$, $u(x_0) > 0$. Then, since u is l.s.c., $u(x) > 0$ in a neighbourhood of x_0. Now $\lim_{n \to \infty} nu$ is, by Theorem 4.4 (3) $\in \mathcal{L}\mathcal{H}_G^*$ and is $+\infty$ on an open set and so everywhere in G. But then $u(x) > 0$ in G. Thus (1).

We may assume that G is a domain. Consider the function u^* defined in $\bar{G}$ by

$$\begin{aligned} u^*(x) &= u(x)/h(x), \quad x \in G \\ &= 0, \quad x \in \partial G. \end{aligned}$$

Then $u^*(x) > -\infty$ in $\bar{G}$ and is l.s.c. there. Suppose, if possible, that u takes negative values in G, and let $k = \inf_G u/h$. Then the value k is attained at some point $x_0 \in G$.

Now the set of functions $\{f/h \mid f \in \mathscr{H}_G\}$ satisfies Axioms 1, 2 and 3, having the same regular sets as the original harmonic structure. Furthermore

$$\frac{f(x)}{h(x)} = \int \frac{f(\xi)}{h(\xi)} \frac{h(\xi)}{h(x)} \sigma_x^B(d\xi),$$

so that the harmonic measure associated with the set of h-harmonic functions is $h(\xi)\,(h(x))^{-1}\,\sigma_x^B$. Also

$$\frac{u(x)}{h(x)} - k \geqslant \int \left(\frac{u(\xi)}{h(\xi)} - k\right) \frac{h(\xi)}{h(x)} \sigma_x^B(d\xi)$$

so that $uh^{-1} - k$ is locally h-hyperharmonic in G and non-negative. By (1) of the theorem, $u = kh$ everywhere in G and, since $k < 0$ and h is strictly positive, we have $\lim_{x \in G,\, x \to \xi} u(x) < 0$ for $\xi \in \partial G$. This is a contradiction; and so (2) is proved.

It is now obvious that a maximum principle for hypoharmonic functions holds, and the statement and proof may be safely left to the reader.

We may now show that $\mathscr{L}\mathscr{H}_G^* = \mathscr{H}_G^*$.

Let B be a regular set such that $\bar{B} \subset G$ and let $u \in \mathscr{L}\mathscr{H}_G^*$. Let, for $x \in B$,

$$U(x) = u(x) - \int f(\xi)\, \sigma_x^B(d\xi)$$

where $f \in \mathscr{C}(\partial B)$ and $f \leqslant u$ in ∂B. Then U is l.s.c. in B, $U(x) > -\infty$ in B and, for $\bar{B}_\alpha(z) \subset B$,

$$\int U(x)\, \sigma_z^{B_\alpha(z)}(dx) = \int u(x)\, \sigma_z^{B_\alpha(z)}(dx) - \int H_f^B(x)\, \sigma_z^{B_\alpha(z)}(dx).$$

The first term on the right hand side is not greater than $u(z)$, since $u \in \mathscr{L}\mathscr{H}_G^*$ and the second term is equal to $H_f^B(z)$ since H_f^B is harmonic in B. Thus $U \in \mathscr{L}\mathscr{H}_B^*$ and

$$\liminf_{x \in B,\, x \to \xi} U(x) \geqslant u(\xi) - f(\xi) \geqslant 0, \qquad \xi \in \partial B.$$

So, by Theorem 4.6 (2), $U(x) \geqslant 0$ in B, and thus

$$u(x) \geqslant \int f(\xi)\, \sigma_x^B(d\xi).$$

Since we are assuming only that $f \leqslant u$, it follows that

$$u(x) \geqslant \int u(\xi)\, \sigma_x^B(d\xi)$$

and so $u \in \mathscr{H}_G^*$.

This done, we turn now to a theorem which plays a crucial role in almost all that follows and certainly in the construction of a solution to the Dirichlet Problem. It is the abstract counterpart of Theorem 2.27 and Theorem 2.30 and states:

THEOREM 4.7. *Suppose that* $u \in \mathscr{H}_G^*$ *and that* $B \subset \bar{B} \subset G$ *is regular. Suppose that* u_B *is defined in* G *by*

$$u_B(x) \begin{cases} = u(x), & x \in G \setminus B \\ = \int u(\xi)\, \sigma_x^B(d\xi), & x \in B. \end{cases}$$

Then (1) $u_B(x) \leqslant u(x), \quad x \in G,$

(2) u_B *is either harmonic in* B *or is* $+\infty$ *in* B,

(3) $u_B \in \mathscr{H}_G^*$.

(4) *Let* $\mathfrak{B}_a$ *be the family of all regular sets containing* a. *Then, for* $a \in G$,

$$u(a) = \sup\{u_B(a) \mid B \in \mathfrak{B}_a\}.$$

Since $u \in \mathscr{H}_G^*$, $u(x) \geqslant u_B(x)$ for $x \in B$ and this gives (1). (2) follows from Theorem 4.3. For (3) it is enough to show that $u_B \in \mathscr{L}\mathscr{H}_G^*$. Given any $x \in G \setminus \partial B$ we shall choose the regular neighbourhood base at x to be those regular sets B_α such that $x \in B_\alpha$ and $\bar{B}_\alpha \subset G \setminus \partial B$. Then, if $x \in G \setminus \bar{B}$ choose $\bar{B}_\alpha \subset G \setminus \bar{B}$ so that, since $u \in \mathscr{H}_G^*$,

$$u_B(x) = u(x) \geqslant \int u(\xi)\, \sigma_x^{B_\alpha}(d\xi) = \int u_B(\xi)\, \sigma_x^{B_\alpha}(d\xi)$$

so that u_B satisfies (3′) at x. Furthermore u_B is (with u) l.s.c. at x and $u_B(x) > -\infty$.

Alternatively, if $x \in B$ then u_B is either harmonic or $+\infty$ in B and so is l.s.c. and $> -\infty$ and thus hyperharmonic in B. It therefore satisfies the conditions for local hyperharmonicity in $G \setminus \partial B$.

It therefore remains to show that u_B satisfies the three conditions for $\mathscr{L}\mathscr{H}_G^*$ at all points in ∂B. Now $u_B(\xi) = u_B(\xi)$ for $\xi \in \partial B$ so $u_B(\xi) > -\infty$. Next, choose $f \in \mathscr{C}(\partial B)$ so that $f(\xi) \leqslant u(\xi)$ for $\xi \in \partial B$. Then, for $x \in B$,

$$u_B(x) \geqslant \int f(\xi)\, u_x^B(d\xi) = H_f^B(x).$$

Now $\lim\limits_{x \to \xi} H_f^B(x) = f(\xi)$ for $\xi \in \partial B$, and so $\liminf\limits_{x \in B,\, x \to \xi} u_B(x) \geqslant f(\xi)$. Hence $\liminf\limits_{x \in B,\, x \to \xi} u_B(x) \geqslant u_B(\xi)$ for $\xi \in \partial B$. When $x \in G \setminus B$, $u_B(x) = u(x)$ and, since u is l.s.c., $\liminf\limits_{x \in G \setminus B,\, x \to \xi} u_B(x) \geqslant u(\xi)$. Thus $\liminf\limits_{x \to \xi} u_B(x) \geqslant u_B(\xi)$ and so u_B is l.s.c. in G.

Finally

$$u_B(\xi) = u(\xi) \geqslant \int u(\eta)\, \sigma_{\xi_1}^{B_\alpha(\xi)}(d\eta) \geqslant \int u_B(\eta)\, \sigma_{\xi}^{B_\alpha(\xi)}(d\eta)$$

so that u_B satisfies condition (3') at ξ. We thus have (3).

We turn to (4). We have $u(a) > u_B(a)$ when $B \in \mathfrak{B}_a$. Also, given $\varepsilon > 0$, there is a $B \in \mathfrak{B}_a$ such that, firstly,

$$u(\xi) > u(a) - \varepsilon \text{ for } \xi \in \partial B$$

and, secondly, by Lemma 4.2,

$$\int \sigma_a^B(d\xi) > 1 - \varepsilon,$$

and thus

$$u_B(a) > (u(a) - \varepsilon)\,(1 - \varepsilon).$$

Since ε is arbitrary, this gives (4).

We note that, if u is superharmonic in G, then u_B is harmonic in B and, of course, superharmonic in G since $u_B \leqslant u$. Conversely, if u_B is harmonic in B, then u is superharmonic in the component containing B since, if it were not, it would be $+\infty$ in ∂B and so $u_B = +\infty$ in B.

Also, if $u \in (-\mathscr{H}_G^*)$, then Theorem 4.7 holds, except that $\geqslant$ replaces $\leqslant$ in (1), $-\infty$ replaces ∞ in (2), $-\mathscr{H}_G^*$ replaces $\mathscr{H}_G^*$ in (3) and 'inf' replaces 'sup' in (4).

§ 4.3. Saturated Sets

Let $G \subset X$ be open. **A set $\mathscr{A} \subset \mathscr{H}_G^*$ is said to be saturated in G if**

(*i*) if $u_1, u_2 \in \mathscr{A}$ then $\inf (u_1, u_2) \in \mathscr{A}$;

(*ii*) if $B \subset \bar{B} \subset G$ is regular and $u \in \mathscr{A}$, then $u_B \in \mathscr{A}$.
(u_B was defined in Theorem 4.7.)

Similarly a set $\mathscr{A} \subset (-\mathscr{H}_G^*)$ **is said to be saturated in G** if $-\mathscr{A}$ is saturated in G.

THEOREM 4.8. *Let $\mathscr{A} \subset \mathscr{H}_G^*$ $(-\mathscr{H}_G^*)$ be saturated in a domain G. Then* $\inf_{\mathscr{A}} u$ $(\sup_{\mathscr{A}} u)$ *is either $+\infty$ everywhere in G, or $-\infty$ everywhere in G or is harmonic in G.*

Let $\mathscr{A} \subset \mathscr{H}_G^*$, and let B be regular and such that $\bar{B} \subset G$. Then $u_B \leqslant u$ for every $u \in \mathscr{A}$ and $\inf_{u \in \mathscr{A}} u_B \leqslant \inf_{\mathscr{A}} u$. But $u_B \in \mathscr{A}$, so $u_B \geqslant \inf_{\mathscr{A}} u$, and hence $\inf_{\mathscr{A}} u_B = \inf_{\mathscr{A}} u$. Furthermore, $\{u_B\}_{u \in \mathscr{A}}$ is a down-directed family, since $\mathscr{A}$ is. Also $u_B \in \mathscr{H}_B$ or is $+\infty$, and so we have a family of functions each either $+\infty$ or harmonic in B. In consequence, the infimum is either $+\infty$ or $-\infty$ or harmonic in B.

Thus the sets in which the infimum takes the value $+\infty$, or takes the value $-\infty$, or is harmonic, are disjoint open sets of which the union is G. Since G is connected, only one of these three sets can be non-void and the result follows.

We turn now to some applications of the concept of saturated set and of the theorem which we have just proved.

The first is that of **greatest harmonic minorant** of a superharmonic function. Suppose that u is superharmonic in G, and consider the set

$$\mathscr{A} = \{f \in (-\mathscr{H}_G^*) \mid f(x) \leqslant u(x),\ x \in G\}.$$

Now $\sup(f_1, f_2) \in \mathscr{A}$ with f_1 and f_2, and, since f_B is subharmonic in G and harmonic in B for every $\bar{B} \subset G$ and $f_B(\xi) = f(\xi) \leqslant u(\xi)$ for $\xi \in \partial B$, we have

$$\liminf_{x \in B,\, x \to \xi} (u(x) - f_B(x)) = \liminf_{x \to \xi} u(x) - \limsup_{x \to \xi} f_B(x) \geqslant u(\xi) - f(\xi) \geqslant 0,$$

and $u - f_B$ is superharmonic in B, we have $f_B(x) \leqslant u(x)$ in B, so that $f_B \in \mathscr{A}$. Thus $\mathscr{A} \subset (-\mathscr{H}_G^*)$ is saturated.

Since u is superharmonic $u \neq +\infty$ in any component of G, and so either $\sup_{\mathscr{A}} f$ is harmonic in a given component of G or takes the value $-\infty$ there.

If $\sup_{\mathscr{A}} f$ is harmonic in G, it is said to be **the greatest harmonic minorant of u in G.** If $\sup_{\mathscr{A}} f$ is not harmonic everywhere in G, then **u is said not to have a harmonic minorant in G.**

Clearly a positive superharmonic function in a set G will always have a greatest harmonic minorant in G. A superharmonic function which has a greatest harmonic minorant 0 in a set G and which is not harmonic in G, is said to be a **potential** in G.

It is clear that we may also speak of the **least harmonic majorant in a set G of a function subharmonic there.**

The next application is to the concept of a **sweep** or **sweep function** of a positive hyperharmonic function over a set.

Let $E \subset X$, and let $u \in \mathscr{H}_X^*$ and $u \geqslant 0$. Let

$$\mathscr{A}_E^u = \{v \in \mathscr{H}_X^* \mid v \geqslant 0 \text{ and } v(x) \geqslant u(x),\ x \in E\},$$

and set

$$^*u^E(x) = \inf\{v(x) \mid v \in \mathscr{A}_E^u\}.$$

There is no reason to suppose that $^*u^E$ is even measurable, and so we may only speak of its upper integral with respect to σ_x^B. We have then

$$\overline{\int} {}^*u^E(\xi)\, \sigma_x^B(d\xi) \leqslant \inf_{v \in \mathscr{A}_E^u} \int v(\xi)\, \sigma_x^B(d\xi)$$

since $\mathscr{A}_E^u$ is a subset of all l.s.c. functions majorising $*u^E$. Since $v \in \mathscr{H}_G^*$ the right hand side above is $\leqslant \inf v(x) = *u^E(x)$, and we thus have

$$*u^E(x) \geqslant \overline{\int} *u^E(\xi)\, \sigma_x^B(d\xi)$$

for all $x \in B$. Now set

$$u^E(x) = \liminf_{y \to x} *u^E(y)$$

and then u^E is l.s.c. in X. Furthermore

$$u^E(x) = \liminf_{y \to x} *u^E(y) \geqslant \liminf_{y \to x} \overline{\int} *u^E(\xi)\, \sigma_y^B(d\xi).$$

Since, by Theorem 4.3, the integral is harmonic and so continuous in B, we have

$$u^E(x) \geqslant \int *u^E(\xi)\, \sigma_x^B(d\xi) \geqslant \overline{\int} u^E(\xi)\, \sigma_x^B(d\xi)$$

since $u^E(\xi) \leqslant *u^E(\xi)$. Also $u^E \geqslant 0$ and thus u^E is superharmonic in X.

We shall say that u^E is the **sweep function** or **sweep** of u over E. In this we will differ somewhat from Brelot's terminology. For him $*u^E$ is the **swept-out** or *balayée* function, while u^E is the **regularised function of** $*u^E$. We shall call $*u^E$ the **pre-sweep,** and the u^E the **sweep.**

The obvious relations $\mathscr{A}_E^{u_1} + \mathscr{A}_E^{u_2} \subset \mathscr{A}_E^{u_1+u_2}$ and $\mathscr{A}_{E_1}^u + \mathscr{A}_{E_2}^u \subset \mathscr{A}_{E_1 \cup E_2}^u$ have the consequences $(u_1+u_2)^E \leqslant u_1^E + u_2^E$ and $u^{E_1 \cup E_2} \leqslant u^{E_1} + u^{E_2}$. Also, when $E_1 \subset E_2$, $\mathscr{A}_{E_1}^u \supset \mathscr{A}_{E_2}^u$ and so $u^{E_1} \leqslant u^{E_2}$. Finally, for $\lambda > 0$, it is easy to see that $(\lambda u)^E = \lambda u^E$.

When we take as the space some domain G, we will obtain the sweep with respect to G of the set $E \subset G$. This we will denote by $(u^E)_G$.

We have already met a simple example of a sweep in § 3.6. The equilibrium potential U_2^γ there is the sweep of the superharmonic function which is equal to 1 everywhere in $\mathbb{R}^p$ over the set K, and the following theorem can be regarded as a generalisation of Theorem 3.29.

THEOREM 4.9.

(1) $0 \leqslant u^E \leqslant *u^E \leqslant u$ *everywhere.*

(2) $*u^E(x) = u(x)$ *when* $x \in E$.

(3) $u^E(x) = u(x)$ *when* $x \in \mathring{E}$.

(4) $u^E(x) = *u^E(x)$ *and is harmonic when* $x \in X \setminus \bar{E}$.

Clearly $u^E \geqslant 0$ and $u^E(x) = \liminf_{y \to x} *u^E(y) \leqslant *u^E(x)$. Also $u \in \mathscr{A}_E^u$ and so $*u^E \leqslant u$. This gives (1).

If $v \in \mathscr{A}_E^u$, then $v(x) \geqslant u(x)$ for $x \in E$, and so

$$*u^E(x) = \inf \{v(x) \mid v \in \mathscr{A}_E^u\} \geqslant u(x) \text{ when } x \in E.$$

This, together with the reverse inequality proved in (1), gives (2).

Suppose that $x \in \mathring{E}$. Then there is a neighbourhood $N \subset E$ of x and, for $y \in N$, $*u^E(y) = u(y)$. Consequently

$$u^E(x) = \liminf_{y \to x} *u^E(y) = \liminf_{y \to x} u(y) = u(x),$$

which gives (3).

Finally, when $v_1, v_2 \in \mathscr{A}_E^u$, then $\inf(v_1, v_2) \in \mathscr{A}_E^u$ and, when $\bar{B} \subset X \setminus \bar{E}$, $v_B(x) \geqslant 0$ for $x \in X \setminus \bar{E}$, since $v(x) \geqslant 0$ there; so $v_B \in \mathscr{A}_E^u$ with v. Thus $\mathscr{A}_E^u$ is a saturated set of hyperharmonic functions in $X \setminus \bar{E}$; and so, by Theorem 4.8, $*u^E$ is harmonic there, since it cannot be $-\infty$ or $+\infty$ in any component of $X \setminus \bar{E}$; the first since $*u^E \geqslant 0$ and the second since $*u^E \leqslant u$. Thus $*u^E$ is continuous in $X \setminus \bar{E}$ and so $u^E = *u^E$, and (4) now follows.

Furthermore, just as Newtonian capacity is associated with U_2^γ, so we can associate a generalised capacity with the sweep.

THEOREM 4.10. *Given $a \in X$ and a continuous potential p on X, the set function ϕ defined on all subsets of X by*

$$\phi(E) = *p^E(a)$$

is a generalised capacity.

First, suppose that $E_1 \subset E_2$. Then $\mathscr{A}_{E_2} \subset \mathscr{A}_{E_1}$ and so $*p^{E_1}(a) \leqslant *p^{E_2}(a)$, and hence $\phi(E_1) \leqslant \phi(E_2)$. Thus ϕ is increasing.

Next, if G is open, $\phi(G) = \sup\{\phi(K) \mid K \subset G$ and K compact$\}$. First, the left-hand side is not less than the right hand side, since ϕ is increasing. Otherwise, $p^K \leqslant p$ and $\{p^K\}_{K \subset G}$ is an up-directed family of superharmonic functions. Hence the right-hand side is superharmonic and, since $p^K(x) = p(x)$ whenever $x \in \mathring{K}$, it follows that $\sup_{K \subset G} p^K(x) \geqslant p(x)$ for $x \in G$, and so $\sup *p^K(x) \geqslant *p^G(x)$. Hence $\sup_{K \subset G} \phi(K) \geqslant \phi(G)$, and the result follows.

Finally $\phi(E) = \inf\{\phi(G) \mid G \supset E$ and G open$\}$. For let $v \geqslant 0$ be superharmonic in X, such that $v \geqslant p$ on E and $*p^E(a) + \varepsilon \geqslant v(a)$. Then, for some open set $G \supset E$, we have

$$v(x) \geqslant p(x)\,(1-\varepsilon) \quad \text{for } x \in G,$$

and so

$$v/(1-\varepsilon) \geqslant *p^G.$$

Then

$$(1-\varepsilon)\, *p^G(a) \leqslant v(a) \leqslant *p^E(a) + \varepsilon$$

so

$$*p^G(a) - *p^E(a) \leqslant \varepsilon(1 + *p^G(a)) \leqslant \varepsilon(1 + p(a)).$$

Thus, given $\varepsilon > 0$, there is an open set $G \supset E$ such that $\phi(G) - \phi(E) < \varepsilon$, and this gives the result.

It follows that $\phi(K) = \inf_{G \supset K} \phi(G)$, and so ϕ is continuous on the right.

Next, ϕ is strongly sub-additive. For let K_1, K_2 be two compacts. If $a \in K_1 \cup K_2$ and, say, $a \in K_1$ then $\phi(K_1 \cup K_2) = p(a) = \phi(K_1)$ and $\phi(K_1 \cap K_2) \leqslant \phi(K_2)$ and strong sub-additivity follows.

Otherwise suppose that $a \in C(K_1 \cup K_2)$ and, for clarity, denote $\phi(E)$ by $\phi_a(E)$. Suppose that v_1 and v_2 are superharmonic and non-negative in X and such that $v_1 \geqslant p$ on K_1 and $v_2 \geqslant p$ on K_2. Then it will be enough to show that, for $a \in X$,

$$D(a) = v_1(a)+v_2(a)-\phi_a(K_1 \cup K_2)-\phi_a(K_1 \cap K_2) \geqslant 0.$$

Since, in this instance, the pre-sweeps are u.s.c. we have, for $b \in \partial\,(K_1 \cup K_2)$,

$$\limsup_{a \in C(K_1 \cup K_2),\, a \to b} (\phi_a(K_1 \cup K_2)+\phi_a(K_1 \cap K_2)$$

$$\leqslant \phi_b(K_1 \cup K_2)+\phi_b(K_1 \cap K_2) \leqslant p(b)+p(b)$$

and so

$$\liminf_{a \in C(K_1 \cup K_2),\, a \to b} D(a) \geqslant 0.$$

Furthermore, D is superharmonic in $\complement\,(K_1 \cup K_2)$ and $D \geqslant -2p$. The function w given by $\inf(D, 0)$ in $C(K_1 \cup K_2)$ and 0 in $K_1 \cup K_2$ is again superharmonic in X and $w \geqslant -2p$. Consequently w is not less than the greatest harmonic minorant of $-2p$ which, since p is a potential, is 0. Thus $w \geqslant 0$ and so $D \geqslant 0$. This now gives the strong sub-additivity.

Thus the restriction of ϕ to compacts is a Choquet capacity and so ϕ is a generalised capacity.

There is also a connection between greatest harmonic minorant and sweep, which we describe in the next theorem.

THEOREM 4.11. *Let u be a positive superharmonic function in X. Then, if G is an open set,*

$$\operatorname*{rest}_{G}\ \inf_{\mathscr{A}} v \tag{4.4}$$

where $\mathscr{A} = \{u^{X \setminus H} \mid H$ open and $\bar{H} \subset G\}$, is the greatest harmonic minorant of u in G.

First, (4.4) *is harmonic in G*. The set $\mathscr{B}$ consisting of all the minima of finite collections of functions in $\mathscr{A}$ and of v_B (defined in Theorem 4.7) for every one of these v mentioned is clearly saturated in G. Consequently $\inf_{\mathscr{B}} v$ is harmonic in G.

Now, if $\bar{H}_1, \bar{H}_2 \subset G$ then $u^{X \setminus H_1 \cup H_2} \leqslant \min\,[u^{X \setminus H_1}, u^{X \setminus H_2}]$. Also, given any regular set B with $\bar{B} \subset G$ and any H with $\bar{H} \subset G$, we can find a like H' such that $\overline{H \cup B} \subset H' \subset \bar{H}' \subset G$. Then

$$u^{X \setminus H'} = (u^{X \setminus H'})_B \leqslant (u^{X \setminus H})_B$$

since $u^{X\setminus H'}$ is harmonic in $H' \supset B$. Thus $\mathscr{A} \subset \mathscr{B}$ and any member of $\mathscr{B}$ is minorised by a member of $\mathscr{A}$, and so $\inf_{\mathscr{B}} v = \inf_{\mathscr{A}} v$, so that $\inf_{\mathscr{A}} v$ is harmonic in G.

Second, suppose $h \in \mathscr{H}_G$ and that $h \leqslant u$ in G. For any $\overline{H} \subset G$, ${}^*u^{X\setminus H}(\xi) = u(\xi)$ for $\xi \in \partial H$ and so $h(\xi) \leqslant {}^*u^{X\setminus H}(\xi)$. Suppose $w \in \mathscr{A}^u_{X\setminus H}$. Then $h(\xi) \leqslant w(\xi)$ for $\xi \in \partial H$ and h is continuous in ∂H, so that

$$\limsup_{x\in H,\, x\to\xi} h(x) \leqslant \limsup_{x\to\xi} h(x)$$

$$= h(\xi) \leqslant w(\xi) = \liminf_{x\to\xi} w(x) \leqslant \liminf_{x\in H,\, x\to\xi} w(x)$$

Consequently,

$$\liminf_{x\in H,\, y\to\xi} (w(x)-h(x)) = \liminf_{x\in H,\, x\to\xi} w(x) - \limsup_{x\in H,\, x\to\xi} h(x) \geqslant 0.$$

Since $w-h \in \mathscr{H}^*_H$ we have $w(x) \geqslant h(x)$ for $x \in H$. Thus ${}^*u^{X\setminus H}(x) \geqslant h(x)$ when $x \in H$; and hence, for $x \in H$,

$$u^{X\setminus H}(x) = \liminf_{y\to x} {}^*u^{X\setminus H}(y) \geqslant \liminf_{y\to x} h(y) = h(x)$$

and this whenever $\overline{H} \subset G$. Thus this inequality holds for $x \in G$ and so $\inf_{\mathscr{A}} v(x) \geqslant h(x)$ for $x \in G$. It follows that (4) is the greatest harmonic minorant of u in G.

The third application is perhaps the central one for this chapter. It is crucial for the construction of the upper and lower solutions for the Dirichlet Problem.

Let G be a relatively compact open set in X and suppose that ∂G is non-void. Let f be a numerical-valued function defined in ∂G. Let

$$\mathscr{O}_f = \{u \in \mathscr{H}^*_G \mid \liminf_{x\in G,\, x\to\xi} u(x) \geqslant f(\xi) \text{ for } \xi \in \partial G\}$$

and

$$\mathscr{U}_f = \{v \in (-\mathscr{H}^*_G) \mid \limsup_{x\in G,\, x\to\xi} v(x) \leqslant f(\xi) \text{ for } \xi \in \partial G\}$$

and define the upper and lower solutions $\overline{H}^G_f$ and $\underline{H}^G_f$ by

$$\overline{H}^G_f(x) = \inf_{\mathscr{O}_f} u(x); \qquad \underline{H}^G_f(x) = \sup_{\mathscr{U}_f} v(x).$$

Now, if $u_1, u_2 \in \mathscr{O}_f$ then, given $\varepsilon > 0$, and $\xi \in \partial G$, there is a neighbourhood N of ξ such that, for $x \in N \cap G$, $u_i(x) \geqslant f(\xi)-\varepsilon$, $i = 1, 2$. Consequently $\min(u_1, u_2)(x) \geqslant f(\xi)-\varepsilon$ so that $\liminf_{x\in G,\, x\to\xi} \min(u_1, u_2)(x) \geqslant f(\xi)$, and so $\min(u_1, u_2)$ belongs to $\mathscr{O}_f$.

Also, if B is regular and $\overline{B} \subset G$, then $\overline{B} \cap \partial G = \phi$, and so

$$\liminf_{x\in G,\, x\to\xi} u_B(x) = \liminf_{x\in G,\, x\to\xi} u(x) \geqslant f(\xi)$$

and thus $u_B \in \mathcal{O}_f$. Thus $\mathcal{O}_f$ is a saturated set and, by Theorem 4.8, $\overline{H}_f^G$ *is either* $+\infty$ *in a given component of* G, *or* $-\infty$ *there, or is harmonic there.*

Similarly $\mathcal{U}_f$ is a saturated set, and so $\underline{H}_f^G$ *is either* $+\infty$ *in a given component of* G, *or* $-\infty$ *there or is harmonic there.*

§ 4.4. More about Solution Functions

THEOREM 4.12. *Let G be a relatively compact open set and let f, g be numerical-valued in* ∂G. *Then*

(1) *If* $f \leqslant g$ *then* $\overline{H}_f^G \leqslant \overline{H}_g^G$ *and* $\underline{H}_f^G \leqslant \underline{H}_g^G$,

(2) $\overline{H}_{\lambda f}^G(x) = \lambda\, \overline{H}_f^G(x)$ *and* $\underline{H}_{\lambda f}^G(x) = \lambda\, \underline{H}_f^G(x)$ *when* $\lambda > 0$,

(3) $\underline{H}_f^G(x) = -\overline{H}_{-f}^G(x)$,

(4) $\underline{H}_{f+g}^G \geqslant \underline{H}_f^G + \underline{H}_g^G$; $\overline{H}_{f+g}^G \leqslant \overline{H}_f^G + \overline{H}_g^G$,

(5) If $\{f_n\}$ *converges increasingly to f then, provided* $\overline{H}_{f_1}^G \geqslant -\infty$, $\{\overline{H}_{f_n}^G\}$ *converges increasingly to* $\overline{H}_f^G$; *if* $\{f_n\}$ *converges decreasingly to f then, provided* $\underline{H}_{f_1}^G < +\infty$, $\{\underline{H}_{f_n}^G\}$ *converges decreasingly to* $\underline{H}_f^G$.

If $f \leqslant g$ then $\mathcal{O}_g \subset \mathcal{O}_f$ and $\mathcal{U}_f \subset \mathcal{U}_g$ and so $\inf_{\mathcal{O}_g} u \geqslant \inf_{\mathcal{O}_f} u$ and $\sup_{\mathcal{U}_g} v \geqslant \sup_{\mathcal{U}_f} v$, which gives (1).

Next, for $\lambda > 0$, $\mathcal{O}_{\lambda f} = \lambda \mathcal{O}_f$ and $\mathcal{U}_{\lambda f} = \lambda \mathcal{U}_f$ so that $\inf_{\mathcal{O}_{\lambda f}} u = \lambda \inf_{\mathcal{O}_f} u$ and $\sup_{\lambda \mathcal{U}_f} v = \lambda \sup_{\mathcal{U}_f} v$ whence (2). Again $\mathcal{U}_f = -\mathcal{O}_{-f}$ from which (3) follows. Next $\mathcal{O}_{f+g} \supset \mathcal{O}_f + \mathcal{O}_g$ and $\mathcal{U}_{f+g} \supset \mathcal{U}_f + \mathcal{U}_g$ and so

$$\overline{H}_{f+g}^G(x) = \inf_{\mathcal{O}_{f+g}} w(x) \leqslant \inf_{w \in \mathcal{O}_f + \mathcal{O}_g} w(x) = \inf \{u_1(x) + u_2(x) \mid u_1 \in \mathcal{O}_f \text{ and } u_2 \in \mathcal{O}_g\} = \inf_{\mathcal{O}_f} u + \inf_{\mathcal{O}_g} u = \overline{H}_f^G(x) + \overline{H}_g^G(x)$$

and there is a similar argument to show the second result in (4) as well.

From (1) $\{\overline{H}_{f_n}^G\}$ is increasing and the result is obviously true if $\lim_{n\to\infty} \overline{H}_{f_n}^G = +\infty$ (and if it is so at one point it will be everywhere in the component containing that point). Also, since $f_n \leqslant f$, $\lim_{n\to\infty} \overline{H}_{f_n}^G(x) \leqslant \overline{H}_f^G(x)$. It therefore remains to show that

$$\lim_{n\to\infty} \overline{H}_{f_n}^G(x) \geqslant \overline{H}_f^G(x) \text{ for } x \in G. \tag{4.5}$$

Let $a \in G$, and choose $\varepsilon > 0$. We can find, for each n, $v_n \in \mathcal{O}_{f_n}$ such that $\liminf_{x \to \xi} v_n(x) \geqslant f_n(\xi)$ and $v_n(a) \leqslant \overline{H}_{f_n}^G(a) + \varepsilon 2^{-n}$. Let

$$w(x) = \lim_{n\to\infty} \overline{H}_{f_n}^G(x) + \sum_{n=1}^{\infty} (v_n(x) - \overline{H}_{f_n}^G(x))$$

Then $w \in \mathscr{H}_G^*$ since $\overline{H}_{f_n}^G \in \mathscr{H}_G^*$ and $\{\overline{H}_{f_n}^G\}$ is increasing and since $v_n - \overline{H}_{f_n}^G$ belongs to $\mathscr{H}_G^*$, is $\geqslant 0$, and so the sum on the right has increasing partial sums, each of which belongs to $\mathscr{H}_G^*$. Next, for each n, $w \geqslant \lim \overline{H}_{f_n}^G + v_n - \overline{H}_{f_n}^G$. Consequently $\liminf_{x \to \xi} w(x) \geqslant \liminf_{x \to \xi} v_n(x) \geqslant f_n(\xi)$ and so $\liminf_{x \to \xi} w(x) \geqslant f(\xi)$ and $w \in \mathcal{O}_f$. Consequently $w \geqslant \overline{H}_f^G$ and $w(a) \leqslant \lim_{n \to \infty} \overline{H}_{f_n}^G(a) + \varepsilon$. This last holds for each $\varepsilon > 0$, so

$$w(a) \leqslant \lim_{n \to \infty} \overline{H}_{f_n}^G(a).$$

Since also $w(a) \geqslant \overline{H}_f^G(a)$ we have $\overline{H}_f^G(a) \leqslant \lim_{n \to \infty} \overline{H}_{f_n}^G(a)$, and since a is arbitrary, this gives the first part of (5).

Furthermore, $\underline{H}_{f_n}^G = -\overline{H}_{(-f_n)}^G$, $\underline{H}_f^G = -\overline{H}_{(-f)}$ and $\{-f_n\}$ converges increasingly to $-f$ when $\{f_n\}$ converges decreasingly to f, so, by the result just proven, the second part of (5) also follows.

So far we have two solution functions to a problem for which we are seeking a unique solution. If, of course, the two solutions coincide in any particular case, then we at least have one candidate for the solution to the Dirichlet Problem.

For any relatively compact open set G and a function f defined in ∂G we say that **f is resolutive for G** if, for all $x \in G$,

$$\underline{H}_f^G(x) = \overline{H}_f^G(x)$$

and both are harmonic in G. When this happens, we denote the common value by $H_f^G(x)$ and say that H_f^G is **the generalised solution to the Dirichlet Problem.**

THEOREM 4.13. *Suppose that f and g are resolutive for G. Then*

(1) *If $f \leqslant g$ then $H_f^G \leqslant H_g^G$.*

(2) *For $\alpha \in \mathbb{R}$, αf is resolutive for G and $H_{\alpha f}^G = \alpha H_f^G$.*

(1) follows immediately from Theorem 4.12 (1)

If $\alpha > 0$, $\overline{H}_{\alpha f}^G = \underline{H}_{\alpha f}^G = \alpha H_f^G$ from Theorem 4.12 (2)

If $\alpha < 0$, $\overline{H}_{\alpha f}^G = -\underline{H}_{(-\alpha)f}^G = -(-\alpha)\,\underline{H}_f^G = \alpha H_f^G$

and $\underline{H}_{\alpha f}^G = -\overline{H}_{(-\alpha)f}^G = -(-\alpha)\,\overline{H}_f^G = \alpha H_f^G$

and so, for $\alpha \in \mathbb{R}$, $\overline{H}_{\alpha f}^G = \overline{H}_{\alpha f}^G$ so that αf is resolutive and $H_{\alpha f}^G = \alpha H_f^G$.

THEOREM 4.14. *Suppose that there is an $h \in \mathscr{H}_G$ such that $\inf_G h > 0$. Then*

$$\underline{H}_f^G(x) \leqslant \overline{H}_f^G(x) \text{ for } x \in G$$

and if there is equality at any point in G there is equality in the component of G containing that point.

We may suppose that G is connected. Let $u \in \mathcal{O}_f$ and $v \in \mathcal{U}_f$. Then $\liminf_{x\to\xi} (u(x)-v(x)) \geqslant \liminf_{x\to\xi} u(x) - \limsup_{x\to\xi} v(x) \geqslant f(\xi)-f(\xi) = 0$. Also $u-v \in \mathcal{H}^*_G$ and so, by Theorem 4.6(2), $u(x)-v(x) \geqslant 0$ for all $x \in G$ and hence $\underline{H}^G_f(x) \leqslant \overline{H}^G_f(x)$ for all $x \in G$.

Also, if there is equality at a point in G then the non-negative harmonic function $\overline{H}^G_f - \underline{H}^G_f$ attains the value 0 at that point and so, by Theorem 4.1, is zero everywhere in G. This deals with the case in which $\overline{H}^G_f$ is harmonic. If $\overline{H}^G_f$ is $+\infty$ then $\underline{H}^G_f$ has the value $+\infty$ at a point and so everywhere and hence we have equality. If $\overline{H}^G_f = -\infty$ then $\underline{H}^G_f = -\infty$.

It will be seen that in the last theorem we have had to introduce an extraneous requirement concerning the existence of a positive harmonic function in G. We shall need this assumption in the next theorem also. Having the next theorem proved, it will be necessary to introduce an Axiom P, to which we have referred earlier. At this time it is worthwhile emphasising what we shall show later, viz. that the existence of a positive harmonic function in G is a **consequence** of Axiom P.

THEOREM 4.15. *Suppose there is an $h \in \mathcal{H}_G$ such that $\inf_G h > 0$. Then*

(1) *If f and g are resolutive for G and $\alpha, \beta \in \mathbb{R}$ then $\alpha f + \beta g$ is resolutive for G and $H^G_{\alpha f+\beta g} = \alpha H^G_f + \beta H^G_g$.*

(2) *If $\{f_n\}$ is a monotonic sequence of functions each resolutive for G and $\{H^G_{f_n}(x_0)\}$ is bounded for some point $x_0 \in G$ then $f = \lim_{n\to\infty} f_n$ is resolutive for G and $H^G_f = \lim_{n\to\infty} H^G_{f_n}$.*

(3) *If $\{f_n\}$ is a uniformly convergent sequence of finite-valued functions each resolutive for G then $f = \lim_{n\to\infty} f_n$ is resolutive. Further, $H^G_{f_n}$ converges to H^G_f uniformly in G provided $\overline{H}^G_1$ is bounded in G.*

From Theorem 4.12 (4)

$$H_f + H_g = \underline{H}_f + \underline{H}_g \leqslant \underline{H}_{f+g} \leqslant \overline{H}_{f+g} \leqslant \overline{H}_f + \overline{H}_g = H_f + H_g$$

so that $\underline{H}_{f+g} = \overline{H}_{f+g}$, $f+g$ is resolutive for G and $H_{f+g} = H_f + H_g$. (1) now follows from Theorem 4.12(2).

Suppose first that $\{f_n\}$ is increasing. By Theorem 4.12(1) $H^G_{f_n} = \underline{H}^G_{f_n} \leqslant \underline{H}^G_f$ and, by 4.12(5), $H^G_{f_n} = \overline{H}^G_{f_n}$ converges increasingly to $\overline{H}^G_f$. Consequently $\underline{H}^G_f = \overline{H}^G_f$ and $\overline{H}^G_f(x_0)$ is finite so H^G_f is harmonic. Thus f is resolutive and $H^G_f = \lim_{n\to\infty} H^G_{f_n}$.

If $\{f_n\}$ is decreasing, then $\{-f_n\}$ is increasing and $H^G_{f_n} = -H^G_{(-f_n)}$ which converges decreasingly to $-H^G_{-f} = H^G_f$. Also $-f$ is resolutive; so f is resolutive. This completes (2).

The function $\overline{H}_1$ is finite. For, since there is a harmonic function in G with positive lower bound, there is one with lower bound $\geqslant 1$. This last is a member of $\mathcal{O}_1$ so not less than $\overline{H}_1$ and is finite.

Now, given $\varepsilon > 0$, we can find n_0 such that, for all $n > n_0$, and for $\xi \in \partial G$

$$f_n(\xi)-\varepsilon < f(\xi) < f_n(\xi)+\varepsilon$$

and thus, by Theorem 4.12(1) and (4), for all $x \in G$,

$$\underline{H}^G_{f_n}(x)+\underline{H}^G_{-\varepsilon}(x) \leqslant \underline{H}^G_f(x) \leqslant \overline{H}^G_f(x) \leqslant \overline{H}^G_{f_n}(x)+\overline{H}^G_{\varepsilon},$$

which gives, since f_n is resolutive for G and using Theorem 4.12(2)

$$H^G_{f_n}(x)-\varepsilon H^G_1(x) \leqslant \underline{H}^G_f(x) \leqslant \overline{H}^G_f(x) \leqslant H^G_{f_n}(x)+\varepsilon H^G_1(x).$$

From this we deduce that

$$0 \leqslant \overline{H}^G_f(x)-\underline{H}^G_f(x) < 2\varepsilon H^G_1(x).$$

Since ε is arbitrary, this shows that the upper and lower solutions coincide; and so f is resolutive.

Also, now

$$H^G_f(x)-\varepsilon\overline{H}^G_1(x) \leqslant H^G_{f_n}(x) \leqslant H^G_f(x)+\varepsilon\overline{H}^G_1(x)$$

which, if $\overline{H}^G_1$ is bounded in G shows that $H^G_{f_n}$ converges uniformly to H^G_f.

Theorem 4.15(1) states that the set of all functions resolutive for G is a linear space over $\mathbb{R}$ and that, for each fixed $x \in G$, $f \to H^G_f(x)$ is a linear functional on this linear space. Theorem 4.13(1) gives, for $f \geqslant 0$, $H^G_f \geqslant \overline{H}^G_0 = H^G_0$ and $H^G_0 = 0$ since 0 is harmonic in G. Thus *$f \to H^G_f(x)$ is a positive linear functional on the linear space of all functions resolutive for G.*

Clearly now, to make further progress it is necessary to know what this linear space consists of. If it is not at least as large as $\mathscr{C}(\partial G)$, we shall have a rather unsatisfactory theory. At this point, therefore, we introduce the Axiom P which, as we shall see, ensures that the linear space contains $\mathscr{C}(\partial G)$ and a good deal more.

AXIOM P. *There is a function superharmonic in X which is everywhere positive and is not harmonic in X.*

Axiom P has the following equivalent form

There is a potential in X.

For, clearly, if there is a potential in X, then it is itself a function superharmonic in X which is everywhere positive and not harmonic in X. Conversely, if u is the function the existence of which is asserted by Axiom P, then u possess a greatest harmonic minorant h since the function 0 is subharmonic and minorises u. Then $p = u-h$ is a potential, since the greatest harmonic minorant of p is 0.

The state of affairs when Axiom P does not hold is somewhat surprising, in that the supply of harmonic functions is then rather thin. We have:

If Axiom P does not hold in X, then all the positive superharmonic functions in X are harmonic and proportional.

In that case, any everywhere-positive superharmonic function in X is harmonic. Suppose that h_1 and h_2 are two non-negative harmonic functions in X. Then $h = (h_1, h_2)$ is superharmonic and non-negative and so is harmonic. Next, h_1-h is harmonic and non-negative and attains the value 0 if h_2-h does not; and so, by Theorem 4.1, $h_1 = h$. Hence $h_1 < h_2$ everywhere.

Thus, given any two distinct non-negative harmonic functions h_1 and h_2 either $h_1 < h_2$ or $h_1 > h_2$. Let $a \in X$ and let $\lambda = h_2(a)/h_1(a)$. Then λh_1 is harmonic and non-negative and so either $\lambda h_1 < h_2$ or $\lambda h_1 > h_2$ or $\lambda h_1 = h_2$; since $\lambda h_1(a) = h_2(a)$ we must have the third possibility.

If p is a potential, then λp $(\lambda > 0)$ is a potential. If u is superharmonic and $0 \leqslant u \leqslant p$, then u is a potential, since the greatest harmonic minorant of u is not greater than that of p but is not less than 0, and is therefore 0. If p_1 and p_2 are potentials, then $\inf(p_1, p_2)$ is a potential by what has just been said. Finally, $\boldsymbol{p_1+p_2}$ **is a potential.** For suppose that p_1+p_2 is not a potential. Then p_1+p_2 is superharmonic, and there is a positive v subharmonic in X such that $v \leqslant p_1+p_2$. Then $v-p_1 \leqslant p_2$ and $v-p_1$ is subharmonic. Since p_2 is a potential, $v-p_1 \leqslant 0$, and so $v \leqslant p_1$. Since p_1 is a potential, $v \leqslant 0$; which is a contradiction.

As we said previously, given a relatively compact open set, it is *a consequence of Axiom P that there is* $h \in \mathscr{H}_G$ *such that* $\inf_G h > 0$. For let p be a potential in X. Then ${}^*p^{X\setminus G}$ is harmonic in G and is strictly positive in $\bar{G}$ and so attains a positive lower bound in G. It is therefore the required harmonic function.

LEMMA 4.16. *There is a continuous potential in X.*

Let p be a potential in X and let G_1 and G_2 be open sets such that $\bar{G}_1 \subset G_2$. Then $p_0 = p^{X\setminus G_2}$ is a potential, since $0 \leqslant p^{X\setminus G_2} \leqslant p$ and p_0 is harmonic in G_2. Next, $p_1 = p_0^{G_1}$ is a potential and is harmonic in $(X\setminus G_1)^0$. Thus p_1, being continuous in $(X\setminus G_1)^0$ and dominated by p_0 which is continuous in G_2 is therefore bounded on any compact set. Thus p_1 is a locally bounded potential which is continuous outside $\bar{G}_1$.

Let now $\bar{G}_1$ be compact, and suppose that $\{B_r\}_{r=1,\ldots,k}$ is a finite covering of $\bar{G}_1$ by regular sets. Let

$$p_2(x) \begin{array}{l} = \int p_1(\xi)\, \sigma_x^{B_1}(d\xi), \quad x \in B_1 \\ = p_1(x), \quad x \in G \setminus B_1, \end{array}$$

so that p_2 is continuous and harmonic in B_1.
Let

$$p_3(x) \begin{array}{l} = \int p_2(\xi)\, \sigma_x^{B_2}(d\xi), \quad x \in B_2 \\ = p_2(x), \quad x \in G \setminus B_2 \end{array}$$

so that p_3 is continuous in $B_1 \cup B_2$.

In general, supposing p_r a potential continuous in $B_1 \cup B_2 \ldots \cup B_{r-1}$, we set

$$p_{r+1}(x) \begin{array}{l} = \int p_r(\xi)\, \sigma_x^{B_r}(d\xi), \quad x \in B_r \\ = p_r(x), \quad x \in G \setminus B_r \end{array}$$

and proceeding in this way, $q = p_{k+1}$ will be the required potential.

LEMMA 4.17. *Suppose that G is a domain. Then there is a continuous potential in X which is not harmonic in G.*

Let G_1 be a relatively compact open set such that $\bar{G}_1 \subset G$. The potential q^{G_1} coincides with q in G_1 and is positive in G. By covering $\bar{G}_1$ with a finite family of regular sets $\{B_r\}$ such that $\cup B_r \subset G$ and repeating the construction employed in the proof of Lemma 4.16, we find a continuous potential $p \leqslant q$ such that $p = q^{G_1}$ outside $\cup B_r$. Hence $p > 0$, and is harmonic outside $\cup B_r$. Moreover, p is not harmonic in G, since, if it were, it would be harmonic in X and so being a potential would be identically zero.

We shall see now that there is a plentiful supply of potentials in X, enough to ensure that continuous functions can be approximated by differences of potentials. This is the content of the following approximation theorem due to Rose-Marie Hervé.

THEOREM 4.18. *A continuous function on the compact set K can be uniformly approximated by the differences of continuous potentials in X.*

Let

$$\mathscr{S} = \{q \mid q = p_1 - p_2;\, p_1, p_2 \text{ continuous potentials in } X\}.$$

Let p_0 be a fixed potential which is positive and continuous in X. Then $\mathscr{A} = \{\operatorname{rest}_K (q/p_0) \mid q \in \mathscr{S}\}$ is a vector subspace of $\mathscr{C}(K)$. Now, if p_1 and

p_2 are continuous potentials, then so is p_1+p_2 and $\inf(p_1, p_2)$. Consequently, since

$$|p_1-p_2| = p_1+p_2-2\inf(p_1, p_2)$$

it follows that $|f| \in \mathscr{A}$ whenever $f \in \mathscr{A}$. Also

$$\sup(f, g) = \tfrac{1}{2}(f+g+|f-g), \inf(f, g) = \tfrac{1}{2}(f+g-|f-g|)$$

and so $\sup(f, g)$ and $\inf(f, g \in \mathscr{A}$ whenever $f, g \in \mathscr{A}$. Also $1 \in \mathscr{A}$.

Finally, $\mathscr{A}$ separates points. For let $x_0 \neq y_0$, $x_0, y_0 \in K$, and let B be a regular set containing x_0 but excluding y_0. Let w be a continuous potential which is not harmonic in B. If $w(x_0) \neq w(y_0)$, then $\mathscr{A}$ separates x_0 and y_0. If $w(x_0) = w(y_0)$ then w_B is a finite continuous potential, w_B is harmonic in B, $w \geqslant w_B > 0$ in B, w is superharmonic in B, and so $w(x_0) \neq w_B(x_0)$.

So $\mathscr{A}$ is a sub-algebra satisfying the requirements of Theorem 1.9. Hence $\bar{\mathscr{A}} = \mathscr{C}(K)$. Now if f is continuous in K so is f/p_0, and this latter can be as closely approximated by a q/p_0 as we please. Since p_0 has a positive lower bound in K, this means that f can be approximated as closely as we please by q and this gives Theorem 4.18.

THEOREM 4.19. *Let G be a relatively compact open set and let $f \in \mathscr{C}(\partial G)$. Then f is resolutive for G.*

Suppose first that $f \in \mathscr{C}(\partial G)$ is such that we can find $u \in \mathscr{H}_G^*$ such that $\lim_{x\to\xi} u(x) = f(\xi)$ for $\xi \in \partial G$. Then $u \in \mathscr{O}_f$, and so $u \geqslant \overline{H}_f^G$. Also

$$\limsup_{x\in G,\, x\to\xi} \overline{H}_f^G(x) \leqslant \limsup_{x\in G,\, x\to\xi} u(x) = f(\xi)$$

and since $\overline{H}_f^G$ is harmonic (it is not $+\infty$, since u, being finite in a neighbourhood, is superharmonic) and so hypoharmonic, it belongs to $\mathscr{U}_f$. But then $\overline{H}_f^G \leqslant \underline{H}_f^G$ and so $\underline{H}_f^G = \overline{H}_f^G$. Thus, in this special case, f is resolutive for G.

If f is not subject to the previous restriction, we can, by the Tietze Extension Theorem (1.5), find a function f^* continuous in $\bar{G}$ which extends f. Then, by Theorem 4.18, there is a sequence $\{f_n^*\}$ with each f_n^* the difference of two continuous potentials, which converges uniformly in $\bar{G}$ to f^*. Now $f_n = \operatorname{rest}_{\partial G} f_n^*$ is the difference of two functions, each of which is of the type we have already shown to be resolutive for G and then, by Theorem 4.15(3), f is resolutive for G.

Thus, the space of all functions resolutive for G contains at least $\mathscr{C}(\partial G)$. So, for each $x \in G$, $H_f^G(x)$ is a positive linear functional on $\mathscr{C}(\partial G)$ and so, by the Riesz Representation Theorem (1.20), there is a unique measure σ_x^G (we extend the notation, previously introduced only for regular sets, to the more general class of relatively compact open sets) such that

$$H_f^G(x) = \int f(\xi)\, \sigma_x^G(d\xi), \qquad f \in \mathscr{C}(\partial G).$$

The measure σ_x^G is said to be **the harmonic measure for G at the point x.** We now argue precisely as in the proof of Theorem 4.3 to obtain

THEOREM 4.20. *Given a relatively compact domain G, the integrability of a function f with respect to σ_x^G is independent of $x \in G$ and, for such an f defined on ∂G,*

$$\int f(\xi)\, \sigma_x^G(d\xi)$$

is harmonic in G.

We may also produce results for $\overline{\int} f(\xi)\, \sigma_x^G(d\xi)$ and $\underline{\int} f(\xi)\, \sigma_x^G(d\xi)$ like those in Theorem 4.3. Furthermore, we have

COROLLARY 4.20.1. *On ∂G the sets of harmonic measure zero are independent of $x \in G$.*

COROLLARY 4.20.2. *Any neighbourhood of any point in ∂G has non-zero σ_x^G-measure.*

We now have a theorem which shows that the class of resolutive functions for G is comfortably as big as we are ever likely to need.

THEOREM 4.21. *A function f defined in ∂G is resolutive for G if, and only if, f is σ_x^G-integrable.*

Suppose first that f is l.s.c. in ∂G. Since ∂G is compact there is, by Theorem 1.7(2), an increasing sequence of continuous functions $\{f_n\}$ converging to f. Since f_1 is continuous and so bounded below in ∂G, $\overline{H}_{f_1}^G > -\infty$ and by Theorem 4.12(5) $\overline{H}_{f_n}^G \to \overline{H}_f^G$. Consequently,

$$\overline{H}_f^G(x) = \lim_{n\to\infty} \int f_n(\xi)\, \sigma_x^G(d\xi) = \int f(\xi)\, \sigma_x^G(d\xi).$$

Also $\overline{H}_{f_n}^G = \underline{H}_{f_n}^G \leqslant \underline{H}_f^G$ by Theorem 4.12(1), and so $\lim_{n\to\infty} H_{f_n}^G(x) \leqslant \underline{H}_f^G(x)$ so that $\overline{H}_f^G(x) \leqslant \underline{H}_f^G(x)$. This gives $\overline{H}_f^G = \underline{H}_f^G$, and so

$$H_f^G(x) = \int f(\xi)\, \sigma_x^G(d\xi).$$

If f is u.s.c., then $-f$ is l.s.c. and $H_f^G(x) = -H_{(-f)}^G(x) = \int f(\xi)\, \sigma_x^G(d\xi)$.

Suppose now that f is any numerical-valued function defined in ∂G. Then

$$\begin{aligned}\overline{\int} f(\xi)\, \sigma_x^G(d\xi) &= \inf\{\textstyle\int \phi(\xi)\, \sigma_x^G(d\xi) \mid \phi \text{ l.s.c. and } \phi \geqslant f\}\\ &= \inf\{H_\phi^G(x) \mid \phi \text{ l.s.c. and } \phi \geqslant f\}\end{aligned}$$

so that

$$\overline{H}_f^G(x) \leqslant \bar{\int} f(\xi)\, \sigma_x^G(d\xi) = \inf_{\phi \geqslant f} H_\phi^G(x) \tag{4.6}$$

Let $\varepsilon > 0$ be given, and choose $x \in G$. Then there is $u \in \mathcal{O}_f$ such that $u(x) \leqslant \overline{H}_f^G(x)+\varepsilon$. Let the function ϕ in ∂G be given by

$$\phi(\xi) = \liminf_{y \in G,\, y \to \xi} u(y).$$

It is l.s.c. and $u \in \mathcal{O}_\phi$, so that $u \geqslant H_\phi^G$. It also majorises f. Thus

$$\overline{H}_f^G(x) \leqslant H_\phi^G(x) \leqslant \overline{H}_f^G(x)+\varepsilon,$$

and so $\inf_{\phi \geqslant f} H_\phi^G(x) \leqslant \overline{H}_f^G(x)$.

This, together with (4.6) gives $\overline{H}_f^G(x) = \bar{\int} f(\xi)\, \sigma_x^G(d\xi)$. Similarly,

$$\underline{H}_f^G(x) = \underline{\int} f(\xi)\, \sigma_x^G(d\xi)$$

and the result now follows.

In particular, we see that all bounded Borel functions in ∂G are resolutive.

If u is a positive superharmonic function in X, and G is a relatively compact open set, then we define the **best harmonic minorant of *u* in *G*** to be H_f^G where $f = \operatorname{rest}_{\partial G} u$.

There is an important connection between best harmonic minorant and sweep, which we bring out in the next theorem.

THEOREM 4.22. *Let u be non-negative and superharmonic in X and let G be a relatively compact open set. Then the best harmonic minorant of u in G is equal to* $u^{X \setminus G}$.

Let $f = \operatorname{rest}_{\partial G} u$. Then we must show that $H_f^G(x) = u^{X \setminus G}(x)$ for $x \in G$. Suppose $w \in \mathcal{O}_f$. Then the function w^*, defined by

$$w^*(x) \begin{cases} = \inf\,(w(x), u(x)), & x \in G \\ = u(x), & x \in X \setminus G, \end{cases}$$

is superharmonic in X. It is clearly so both in G and $X \setminus \bar{G}$; so suppose that $\xi \in \partial G$. Then

$$\liminf_{x \to \xi} w^*(x) = \inf\,[\liminf_{x \to \xi} w(x), \liminf_{x \to \xi} u(x)] = u(\xi) = w^*(\xi)$$

so that w^* is l.s.c. in ∂G.

Next, let B be a regular set containing ξ. Then

$$w^*(\xi) = u(\xi) \geqslant \int u(\eta)\, \sigma_\xi^B(d\eta) \geqslant \int w^*(\eta)\, \sigma_\xi^B(d\eta)$$

so that w^* is superharmonic in X.

Since $w^* \geqslant 0$ and $w^*(x) = u(x)$ for $x \in X \setminus G$ it majorises $^*u^{X\setminus G}$ and so $u^{X\setminus G}$. This holds for any $w \in \mathcal{O}_f$ and so

$$H_f^G(x) \geqslant u^{X\setminus G}, \quad x \in G. \tag{4.7}$$

Also $\liminf_{x \in G,\, x \to \xi} u^{X\setminus G}(x) = \liminf_{x \in G,\, x \to \xi} u(x) = u(\xi)$, $\xi \in \partial G$, since $^*u^{X\setminus G}(x) = u^{X\setminus G}(x)$ for $x \in G$ and $^*u^{X\setminus G}(\xi) = u(\xi)$. Now $u^{X\setminus G}$ belongs to $\mathscr{H}_G^*$ and so $u^{X\setminus G}(x) \geqslant H_f^G(x)$ for $x \in G$; and this, together with (4.7) gives the result.

§ 4.5. Behaviour at the Boundary

The solution function H_f^G has been developed in order to solve the Dirichlet Problem. Although the boundary data has played a determining role in the construction of the upper and lower solution functions, nothing has been said so far about how H_f^G matches up with f on the boundary.

Given a relatively compact open set G, we shall say that a point $\xi \in \partial G$ is **regular for G** if, for every $f \in \mathscr{C}(\partial G)$,

$$\lim_{x \in G,\, x \to \xi} H_f^G(x) = f(\xi).$$

It is then to be noted that, if X is compact, then the boundary data can form the basis for a Dirichlet Problem for G and for a Dirichlet Problem for $X \setminus G$; and then it is clear that regular for G and regular for $X \setminus G$ are by no means the same thing. Indeed, it may well be the case that $\xi \in \partial G$ could be regular for G and irregular for $X \setminus G$.

THEOREM 4.23. *G is regular if, and only if, it is connected and all its boundary points are regular for G.*

If every point of ∂G is regular for G, then $H_f^G \in \mathscr{H}_G$ and $\lim_{x \in G,\, x \to \xi} H_f^G(x) = f(\xi)$ for all $\xi \in \partial G$ and all $f \in \mathscr{C}(\partial G)$. Furthermore H_f^G is unique. For suppose $h_1, h_2 \in \mathscr{H}_G$ and $\lim_{x \to \xi} h_i(\xi) = f(\xi)$ for $\xi \in \partial G$. Then, for $\xi \in \partial G$, $\lim_{x \to \xi} (h_1(x) - h_2(x)) = 0$; and so, by Theorem 4.6(2) applied to $h_1 - h_2$ and $h_2 - h_1$ in turn, we have $h_1 = h_2$. Also, when $f \geqslant 0$, $H_f^G \geqslant 0$. Thus G satisfies all the requirements for a regular set.

Conversely, if G is regular, then a unique H_f^G (in the sense of H_f^B heretofore) exists, which matches up on ∂G and this must be H_f^G (in the new sense). So every point of ∂G is regular for G.

In fact, whether G is regular or not, we can formulate a uniqueness result. *If $f \in \mathscr{C}(\partial G)$ is given and if there is a function h solving the Dirichlet*

Problem for this f, that is, if

$$\lim_{x \in G,\, x \to \xi} h(x) = f(\xi), \quad \xi \in \partial G$$

then h is unique.

For $h \in \mathcal{O}_f^G \cap \mathcal{U}_f^G$ and so $h \geqslant H_f^G \geqslant h$ in G and so $h = H_f^G$. We shall see later (Theorem 4.38) that a much stronger uniqueness result holds.

THEOREM 4.24. *If f is defined in ∂G and $\xi \in \partial G$ is regular for G then*

$$\limsup_{x \in G,\, x \to \xi} \overline{H}_f^G(x) \leqslant \limsup_{\eta \in \partial G,\, \eta \to \xi} f(\eta), \textit{ provided } \overline{H}_{f^-}^G < +\infty$$

$$\liminf_{x \in G,\, x \to \xi} \underline{H}_f^G(x) \geqslant \liminf_{\eta \in \partial G,\, \eta \to \xi} f(\eta), \textit{ provided } \underline{H}_{f^+}^G < +\infty$$

If in the first inequality the right-hand side is $+\infty$, there is nothing to prove. Suppose then that f is bounded above in ∂G, let $\lambda > \limsup_{\eta \in \partial G,\, \eta \to \xi} f(\eta)$ and let $g \in \mathscr{C}(\partial G)$ be such that $g(\xi) < \lambda$ and $g(\eta) \geqslant f(\eta)$ on $\mathscr{C}(\partial G)$. Then $\overline{H}_f^G \leqslant \overline{H}_g^G$ so

$$\limsup_{x \in G,\, x \to \xi} \overline{H}_f^G(x) \leqslant \limsup_{x \in G,\, x \to \xi} H_g^G(x) = g(\xi) < \lambda.$$

Since this holds for all such λ, we have the inequality in this case.

If f be not bounded above in ∂G let $f_n = \min\,[n, f]$ so that f_n, for each n, is bounded above in ∂G. Then

$$\limsup_{x \in G,\, x \to \xi} \overline{H}_{f_n}^G(x) \leqslant \limsup_{\eta \in \partial G,\, \eta \to \xi} f_n(\eta) \leqslant \limsup_{\eta \in \partial G,\, \eta \to \xi} f(\eta)$$

Since $\overline{H}_{f_1}^G \geqslant -\overline{H}_{f^-}^G > -\infty$, $\overline{H}_{f_n}^G \to \overline{H}_f^G$ increasingly, by Theorem 4.12(5); and so

$$\limsup_{x \in G,\, x \to \xi} \overline{H}_f^G(x) \leqslant \limsup_{\eta \in \partial G,\, \eta \to \xi} f(\xi).$$

If we replace f in the first result by $-f$ this gives the second result.

COROLLARY 4.24.1. *For any bounded and resolutive function f which is continuous at a point $\xi \in \partial G$, regular for G, we have*

$$\lim_{x \in G,\, x \to \xi} H_f^G(x) = f(\xi).$$

This is immediate from Theorem 4.24.

COROLLARY 4.24.2. *Let v be a non-negative and superharmonic function in X. Then, if $\xi \in \partial G$ is regular for G,*

$$v^{X \setminus G}(\xi) = v(\xi).$$

Let $f = \operatorname{rest}_{\partial G} v$. Then, by Theorem 4.22, $H_f^G(x) = v^{X \setminus G}(x)$ for $x \in G$. Also, by the second inequality in Theorem 4.24, $\liminf_{x \in G, x \to \xi} H_f^G(x) \geqslant f(\xi) = v(\xi)$ so that $\liminf_{x \in G, x \to \xi} v^{X \setminus G}(x) \geqslant v(\xi)$. Also $*v^{X \setminus G}(x) = v(x)$ for $x \in X \setminus G$ so

$$v^{X \setminus G}(\xi) = \liminf_{x \to \xi} *v^{X \setminus G}(x) \geqslant v(\xi).$$

Since $v^{X \setminus G} \leqslant v$ the equality now follows.

THEOREM 4.25. *A point $\xi \in \partial G$ is regular for G if, and only if, for every continuous potential p in X we have*

$$\lim_{x \in G, x \to \xi} H_p^G(x) = p^{X \setminus G}(\xi) = p(\xi).$$

The necessity for the condition is a consequence of Corollary 4.24.2. Suppose then that the equality holds. Since $\operatorname{rest}_{\partial G} p \in \mathscr{C}(\partial G)$ it is resolutive for G and so

$$\overline{H}_p^G = \underline{H}_p^G = H_p^G$$

(where, by an abuse of notation that we shall find convenient frequently in the sequel, we write p instead of, as we properly should, $\operatorname{rest}_{\partial G} p$). Now $p^{X \setminus G}(x) = H_p^G(x)$ for $x \in G$, by Theorem 4.22. Then

$$p(\xi) = \lim_{x \to \xi} p(x) = \limsup_{x \in G, x \to \xi} p^{X \setminus G}(x) = \limsup_{x \in G, x \to \xi} H_p^G(x)$$

$$\geqslant \liminf_{x \in G, x \to \xi} H_p^G(x) \geqslant \liminf_{x \to \xi} p^{X \setminus G}(x) = p^{X \setminus G}(\xi)$$

(since $p^{X \setminus G}$ is l.s.c.) and this last has the value $p(\xi)$. Hence $\lim_{x \in G, x \to \xi} H_p^G(x)$ exists and has the value $p(\xi)$.

Now, given $f \in \mathscr{C}(\partial G)$ there is, by Theorem 4.18, a sequence $\{q_n\}$ of differences of continuous potentials such that $\operatorname{rest}_{\partial G} q_n$ converges uniformly to f. Then, by Theorem 4.15(3), $H_{q_n}^G(x)$ converges uniformly to $H_f^G(x)$ in G. (The requirement that $\overline{H}_1^G$ be bounded is met here, since we take G' open and such that $\overline{G} \subset G'$. There is a harmonic function h in G' with $\inf_{G'} h > 0$, and so there is a bounded harmonic function in G which is not less than 1 everywhere. This dominates H_1^G, so this last is bounded) and so

$$\lim_{x \to \xi} H_f^G(x) = f(\xi).$$

Consequently ξ is regular for G.

COROLLARY 4.25.2. *If $G_1 \subset G_2$ and $\xi \in \partial G_1 \cap \partial G_2$, then, if ξ is regular for G_2, it is regular for G_1.*

We have $p^{X \setminus G_1} \geqslant p^{X \setminus G_2}$ and thus, for any continuous potential p, $p(\xi) \geqslant p^{X \setminus G_1}(\xi) \geqslant p^{X \setminus G_2}(\xi) = p(\xi)$ so that $p^{X \setminus G_1}(\xi) = p(\xi)$ and then ξ is regular for G_1.

In fact, regularity of a point has a local character as the following theorem shows.

THEOREM 4.26. *If $\xi \in \partial G$ is regular for G, it is regular for $G \cap N$ where N is an open neighbourhood of ξ; and conversely, if ξ is regular for $G \cap N$ for a neighbourhood N of ξ it is regular for G.*

Since $G \cap N \subset G$, the first part of the theorem follows immediately from Corollary 4.25.2.

Conversely, let p be a continuous potential in X, and let N be a neighbourhood of ξ. Suppose that ξ is regular for $G \cap N$.

Define p' in $\bar{G}$ by

$$p'(\xi) = p(\xi), \quad \xi \in \partial G$$

$$p'(x) = H_p^G(x), \quad x \in G$$

Then p', being a bounded Borel function in $\partial(G \cap N)$ is resolutive for $G \cap N$. Also $H_p^G(x) = H_{p'}^{G \cap N}(x)$ for $x \in G \cap N$.

For, let $v \in \mathcal{O}_p^G$ (where again by an abuse of notation, p replaces $\operatorname{rest}_{\partial G} p$). Then $v \geqslant H_p^G$ in G and $\liminf_{x \in G, x \to \xi} v(x) \geqslant p(\xi)$ for $\xi \in \partial G$. Also

$$\liminf_{x \in G \cap N, x \to \eta} v(x) \geqslant p'(\eta), \quad \eta \in \partial G \cap N.$$

Thus $v \in \mathcal{O}_{p'}^{G \cap N}$, and so $v(x) \geqslant H_{p'}^{G \cap N}(x), \quad x \in G \cap N$.

Since this holds for each $v \in \mathcal{O}_p^G$, we have

$$H_p^G(x) \geqslant H_{p'}^{G \cap N}(x), \quad x \in G \cap N.$$

On the other hand, suppose $w \in \mathcal{O}_{p'}^{G \cap N}$. Then define w' by

$$w'(x) = \begin{cases} \inf(v(x), w(x)), & x \in G \cap N \\ v(x), & x \in G \setminus G \cap N \end{cases}$$

Then w' is superharmonic in G. Furthermore $\liminf_{x \in G, x \to \xi} w'(x) \geqslant p(\xi)$ for $\xi \in \partial G$ so $w' \in \mathcal{O}_p^G$ and hence $w'(x) \geqslant H_p^G(x)$ for $x \in G$. Hence $w(x) \geqslant H_p^G(x)$ for $x \in G \cap N$, and so

$$H_{p'}^{G \cap N}(x) \geqslant H_p^G(x), \quad x \in G \cap N,$$

and this gives the required result.

Now, since ξ is regular for $G \cap N$, $\lim_{x \in G \cap N, x \to \xi} H_{p'}^{G \cap N}(x) = p'(\xi) = p(\xi)$ by Corollary 4.24.2 and Theorem 4.22. Then, by the equality above, $\lim_{x \in G, x \to \xi} H_p^G(x) = p(\xi)$, and so ξ is regular for G.

THEOREM 4.27. *Let p be a potential in X which is continuous at a point $\xi \in \partial G$. For ξ to be regular for G, it is necessary and sufficient that for every neighbourhood N of ξ*

$$p^{N \cap (X \setminus G)}(\xi) = p(\xi).$$

Let $K_1 = N \cap (X \setminus G)$. We may suppose that N is compact, since X is locally compact. Then K_1 is compact. Suppose that ξ is regular for G. Let M be a relatively compact open set such that $N \subset M$ and let $M_1 = M \setminus K_1$. Define g on ∂M_1 by

$$g(x) \begin{cases} = p(x), & x \in \partial K_1 \\ = 0, & x \in \partial M. \end{cases}$$

Then $(p^{K_1})_M = H_p^{M_1} \geqslant \overline{H}_g^{M_1} \geqslant \underline{H}_g^{M_1}$. Now, $\mathring{N} \cap M_1 \subset G$ so ξ is regular for $\mathring{N} \cap M_1$, and hence is regular for M_1. Consequently, by Theorem 4.24,

$$\liminf_{x \in M_1, x \to \xi} H_g^{M_1}(x) \geqslant \liminf_{\eta \in \partial K_1, \eta \to \xi} g(\eta) = p(\xi)$$

and so

$$\liminf_{x \to \xi} (p^{K_1})_M(x) = (p^{K_1})_M(\xi) \geqslant p(\xi).$$

Since also $(p^{K_1})_M \leqslant p^{K_1}$ we have $p^{K_1}(\xi) \geqslant p(\xi)$. Thus, finally $p^{K_1}(\xi) = p(\xi)$.

Conversely, suppose, if possible, that ξ be irregular for G while the equality holds. Then, by Theorem 4.25, there is a continuous potential q such that $q^{X \setminus G}(\xi) < q(\xi)$. Choose $\lambda > 0$ so that

$$q^{X \setminus G}(\xi) < \lambda p(\xi) < q(\xi).$$

Then there is a compact neighbourhood N of ξ such that $\lambda p(x) < q(x)$ for $x \in N$. So

$$(\lambda p)^{N \cap (X \setminus G)}(\xi) \leqslant q^{N \cap (X \setminus G)}(\xi) \leqslant q^{X \setminus G}(\xi) < \lambda p(\xi)$$

and hence $p^{N \cap (X \setminus G)}(\xi) < p(\xi)$; which is a contradiction. Hence ξ is regular for G.

§ 4.6. Polar Sets and Thin Sets

We turn next to the question of the size of the subset of ∂G of all irregular points. The measure of size which is peculiarly adapted to the problem is the concept of polar set. We shall see later that for the Laplace case a set is polar if, and only if, it is of zero 2-capacity. In this more general context the concept of capacity is difficult to formulate, since there is no obvious kernel to hand.

A set E is said to be **polar in the open set $G \subset X$,** if there is in G a superharmonic function $v \geqslant 0$ (or, equivalently, a potential) which takes the value $+\infty$ at least in $E \cap G$. The superharmonic function featuring in the above definition will be said to be **associated** with E. Furthermore, we shall say that a property holds **quasi-everywhere in G,** if it holds except possibly in a set which is polar in G.

It is clear that a set is polar in G if, and only if, it is polar in every component of G.

THEOREM 4.28. *Let G be a relatively compact open set, and suppose that E is polar in X. Then*

(1) $\sigma_x^G(E) = 0$;

(2) *an open set is not polar in X;*

(3) *any countable union of sets polar in X is polar in X;*

(4) *if E is a closed set polar in G and G is connected, then $G \setminus E$ is connected.*

(5) *Let E be a closed set which is polar in X. A superharmonic function v in $X \setminus E$, which is bounded below in every compact subset of X, can be uniquely extended into E to make a function superharmonic in X.*

Let χ_E be the indicator function for $E \cap \partial G$ in ∂G. Then

$$\overline{H}_{\chi_E}^G(x) = \int \chi_E(x)\, \sigma_x^G(d\xi).$$

Let $a \in G$ and let v be associated with $E \cap \partial G$ and be such that $v(a) < +\infty$. (This is possible, since, if $v(a) = +\infty$, we let B be a regular set such that $a \in B \subset \overline{B} \subset G$ and consider v_B.) Then, for any integer n,

$$\liminf_{x \in G,\, x \to \xi} \frac{1}{n}\, v(x) \geqslant \chi_E(\xi), \quad \xi \in \partial G$$

and so $\overline{H}_{\chi_E}^G(a) \leqslant \dfrac{1}{n}\, v(a)$ for every n. Hence $\overline{H}_{\chi_E}^G(a) = 0$ so that $\sigma_a^G(E) = 0$. This gives (1).

A function in $\mathscr{H}_X^*$ which takes the value $+\infty$ on an open set is not superharmonic. There is therefore no function associated with an open set. This gives (2)

Let $\{E_n\}$ be a sequence of sets each of which is polar in X. Let $E = \bigcup_{n=1}^{\infty} E_n$. Let $a \in X$, and let B be a regular set containing a. Let v_n be associated with E_n and be such that

$$\int v_n(\xi)\, \sigma_a^B(d\xi) < n^{-2}.$$

(This is possible, since the integral must be finite for otherwise v_n would not be superharmonic, and then multiplication by an appropriate constant gives the inequality.) Let

$$w = \sum_{n=1}^{\infty} v_n.$$

Then $w \in \mathscr{H}_X^*$ and $\int w(\xi)\, \sigma_a^B(d\xi) < +\infty$, so that w is superharmonic. Also, if $a \in E$, then $w(a) > v_n(a) = +\infty$; so w is associated with E, and so E is polar. This gives (3).

Suppose, if possible, that $G \setminus E = G_1 \cup G_2$ with $G_1 \neq \phi$, $G_2 \neq \phi$, $G_1 \cap G_2 = \phi$ and G_1, G_2 open. Let w be associated with E. Then the function w', which is $+\infty$ on G_2 and w on G_1, is hyperharmonic in $G \setminus E$, is $+\infty$ in an open set but not identically $+\infty$ in G. Consequently G is not connected—a contradiction. Thus (4).

Again let w be associated with E. Let $\{v_n\}$ be defined by

$$v_n(x) \begin{array}{l} = +\infty, \quad x \in E, \\ \\ = v(x) + n^{-1} w(x), \quad x \in X \setminus E. \end{array}$$

Then $\{v_n\}$ is a decreasing sequence of functions, each of which is superharmonic in X and is bounded below on every compact; so, setting $u = \lim_{n\to\infty} v_n$, we have

$$u(a) = \lim_{n\to\infty} v_n(a) \geqslant \lim_{n\to\infty} \int v_n(\xi)\, \sigma_x^B(d\xi) = \int u(\xi)\, \sigma_a^B(d\xi)$$

for each regular B. Also

$$\begin{aligned} \hat{u}(a) = \liminf_{x\to a} u(x) &\geqslant \liminf_{x\to a} \int u(\xi)\, \sigma_a^B(d\xi) \\ &= \int u(\xi)\, \sigma_a^B(d\xi) \geqslant \int \hat{u}(\xi)\, \sigma_a^B(d\xi), \end{aligned}$$

where the second equality is a consequence of Theorem 4.3. Thus $\hat{u}$ is superharmonic in X.

Also, for $a \in X \setminus E$, $u(a) = v(a)$ whenever $w(a)$ is finite and, since then u and v agree except on a polar set which is of harmonic measure zero, we have, for $a \in B \subset X \setminus E$,

$$\int u(\xi)\, \sigma_a^B(d\xi) = \int v(\xi)\, \sigma_a^B(d\xi).$$

Then, by Theorem 4.7(4), $\hat{u}(a) \geqslant v(a)$, and since $u(\xi) \geqslant \hat{u}(\xi)$ we have $v(a) \geqslant \hat{u}(a)$ and thus $\hat{u}(a) = v(a)$.

Thus $\hat{u}$ is the required extension. Furthermore, $\hat{u}$ is unique since, again by Theorem 4.7(4),

$$\hat{u}(a) = \sup_{\mathscr{B}_a} \hat{u}_B(a)$$

and the right hand side is unaffected by the values of $\hat{u}$ in a polar set.

We note that if E is polar in X then $X \setminus E$ is dense in X. This follows readily from (2).

THEOREM 4.29. *Suppose that E is polar in X, and that $v > 0$ is superharmonic in X. Then ${}^*v^E = 0$ quasi-everywhere, and $v^E = 0$ everywhere.*

Conversely, if for one positive superharmonic function v we have $v^E = 0$ everywhere, or if there is an $a \in X$ such that $v^E(a) = 0$ then E is a polar set.

Suppose E is polar, and let w be an associated function. Suppose $a \in X$ is such that $w(a) < +\infty$. Then $\lambda w(x) \geqslant v(x)$ for $x \in E$ and any $\lambda > 0$, and so ${}^*v^E(a) = 0$. Hence ${}^*v^E = 0$ quasi-everywhere, and so on a set dense in X. Hence $v^E = 0$ everywhere.

Conversely, suppose that ${}^*v^E(a) = 0$ for some $a \in X$. Then there is a function $v_n > 0$ superharmonic in X such that $v_n \geqslant v$ on E and $v_n(a) < n^{-2}$. Then $\sum_1^\infty v_n$ is superharmonic and positive and is $+\infty$ on E, so that E is polar.

Now suppose that $v^E = 0$ at a point and, being superharmonic and non-negative, everywhere in X. Now, for regular B containing a,

$$v^E(a) = \liminf_{x \to a} {}^*v^E(x) \geqslant \liminf_{x \to a} \overline{\int} {}^*v^E(\xi)\, \sigma_x^B(d\xi) = \overline{\int} {}^*v^E(\xi)\, \sigma_a^B(d\xi)$$

and so the right-hand member is zero. Consequently ${}^*v^E(\xi) = 0$ everywhere in ∂B except at a set of zero σ_a^B-measure and so, certainly, ${}^*v^E(a) = 0$ for some point a, and hence E is polar.

Zaremba's example to show that the Dirichlet Problem is not always soluble was a punctured ball and the removed centre of the ball is, in fact, an irregular point. In a sense, it is an unsatisfactory example, in that the irregular point is isolated from the rest of the boundary. But Lebesgue, in 1913, gave an example of a region in which the irregular point was attached to the rest of the boundary, and is indeed the tip of a very sharp spine driven into the region from outside. The complement of the region is therefore 'thin', in an obvious geometrical sense, in the neighbourhood of this point. We shall now see that something of this sort is generally true for irregular points.

A set $E \subset X$ is said to be **thin at a point $a \notin E$**, if either $a \notin \bar{E}$ or if $a \in \bar{E}$, and there is a superharmonic function $v \geqslant 0$ in X such that

$$\liminf_{x \in E,\, x \to a} v(x) > v(a).$$

A set E is said to be **thin at $a \in E$** if $\{a\}$ is polar and $E \setminus \{a\}$ is thin at a.

It is clear that if a set E is not thin at a, then $a \in \bar{E}$ and, for any superharmonic function $v \geqslant 0$,

$$\liminf_{x \in E,\, x \to a} v(x) = v(a).$$

Also a set E is thin at a if, and only if, for every domain G containing a, $E \cap G$ is thin at a in G (thin in G means that we ask for a function superharmonic in G in the requirement).

THEOREM 4.30. *Let $v > 0$ be superharmonic in X and continuous at $a \in X$. In order that a set E not containing a be thin at a it is necessary and sufficient that there be a neighbourhood N of a such that*

$$*v^{E \cap N}(a) < v(a).$$

If E is thin at a and $a \notin \bar{E}$, we may choose N so that $E \cap N = \phi$, and then the condition is obvious. If $a \in \bar{E}$, there is a superharmonic function $w > 0$ such that $\liminf_{x \in E,\, x \to a} w(x) > w(a)$.

Choose $\lambda > 0$ so that $\lambda v(a)$ lies strictly between the two members of this last inequality. Then there is a neighbourhood N of a such that, for $x \in E \cap N$, $w(x) > \lambda v(x)$. Hence $w > *(\lambda v)^{E \cap N}$ and thus

$$*v^{E \cap N}(a) \leqslant w(a)/\lambda < v(a),$$

which is the required condition.

Conversely, suppose the condition holds for some neighbourhood N of a. Then there is a superharmonic function $w > 0$ such that $w \geqslant v$ in $E \cap N$ and $w(a) < v(a)$. Hence, for $a \in \bar{E}$,

$$\liminf_{x \in E,\, x \to a} w(x) \geqslant v(a) > w(a)$$

and so E is thin at a.

THEOREM 4.31. *Let G be a relatively compact open set. If $X \setminus G$ is thin at $a \in \partial G$, then a is irregular for G.*

$\{a\}$ is polar and $X \setminus G \setminus \{a\}$ is thin at a. Then, by Theorem 4.30, there is a neighbourhood N of a and a positive superharmonic function v such that v is continuous at a and, setting $E = X \setminus G \setminus \{a\}$,

$$v^{E \cap N}(a) < v(a).$$

Set $F = (X \setminus G) \cap N$ so that $F = (E \cap N) \cup \{a\}$. Now

$$v^F(a) \leqslant v^{E \cap N}(a) + v^{\{a\}}(a)$$

and $v^{\{a\}}(a) = 0$, since $\{a\}$ is polar. Also $v^F(a) \geqslant v^{E \cap N}(a)$, and thus $v^{E \cap N}(a) = v^F(a)$. So $v^F(a) < v(a)$; and this, by Theorem 4.27, shows that a is irregular for G.

We now turn to a result about the incidence of thin points in a given set; but to get a satisfactory answer we need here to assume our last axiom, Axiom D, and that X has a countable base.

Let us, to motivate Axiom D, return for a moment to the Laplace case. Given a Newtonian potential V_μ and a superharmonic function

$v > 0$ such that $v(x) \geqslant v_\mu(x)$ for $x \in \text{supp } \mu$, we have V_μ harmonic outside supp μ. Thus, if $G = \mathbb{R}^p \setminus \text{supp } \mu$, we have $v - V_\mu$ superharmonic in G and $v(\xi) - V_\mu(\xi) \geqslant 0$ for $\xi \in \partial G$. If also it were the case that $\liminf_{x \in G, x \to \xi} (v(x) - V_\mu(x)) \geqslant 0$, it would follow by Theorem 2.4 that

$$v(x) \geqslant V_\mu(x), \quad x \in G.$$

Thus, in certain circumstances, **if a superharmonic function dominates a potential on its support, it dominates it off the support also.** This is the Domination Principle, and we import it into the general case by assuming

AXIOM D. *If p is a potential in X which is locally bounded in X and harmonic in an open set G, and q is a potential such that $q \geqslant p$ in $X \setminus G$, then $q \geqslant p$ in G also.*

We have defined, for a function positive and superharmonic in X its greatest harmonic minorant and its best harmonic minorant in a relatively compact open set G. It is clear that, in general, the best harmonic minorant does not exceed the greatest harmonic minorant. But, if we assume Axiom D, we then have

THEOREM 4.32. *Let $u \geqslant 0$ be superharmonic and locally bounded in X, and let G be a relatively compact open set. Then the greatest and best harmonic minorants for u in G coincide.*

It will be enough to prove the result for a locally bounded potential.

By Theorem 4.11, the greatest harmonic minorant for p in G is $\operatorname{rest}_G v$, where

$$v(x) = \inf_{w \in \mathscr{A}} w(x); \qquad \mathscr{A} = \{p^{X \setminus H} \mid H \text{ open and } \bar{H} \subset G\}.$$

Now v is nearly superharmonic, that is, $v(x) \geqslant \int v(\xi)\, \sigma_x^B(d\xi)$, but is not necessarily l.s.c. (the proof is by an argument similar to that for the sweep on p. 130). Also since $p^{X \setminus H} \leqslant p$ we have $v \leqslant p$, and hence $\hat{v}(x) = \liminf_{y \to x} v(y) \leqslant p(x)$. Since $\hat{v}$ is superharmonic, it is also a potential. Furthermore, $v(x) = \hat{v}(x)$ for $x \in X \setminus G$.

Let $w \in \mathscr{A}^{\hat{v}}_{X \setminus G}$. Then $w_1 = \inf(p, w)$ is a potential and $w_1 \geqslant \hat{v}$ in $X \setminus G$. Then, by Axiom D, $w_1 \geqslant \hat{v}$ in G, and so $w \geqslant \hat{v}$ in X; and hence ${}^*v^{X \setminus G}(y) \geqslant \hat{v}$ in X. Consequently

$$v^{X \setminus G}(x) = \liminf_{y \to x} {}^*v^{X \setminus G}(y) \geqslant \hat{v}(x), \quad x \in X.$$

But $v^{X \setminus G}(x) = H_v^G(x)$ for $x \in G$, and so $\operatorname{rest}_G v^{X \setminus G}$ is the best harmonic minorant for v in G, and the theorem now follows.

Let us repeat that, from now on, *we assume Axiom D and that X has a countable base of neighbourhoods.*

Given any function f in X, we define its **regulariser** $\hat{f}$ by

$$\hat{f}(a) = \liminf_{y \to a} f(x).$$

Then $\hat{f}$ is l.s.c. in X, and we have

THEOREM 4.33 (*Choquet's Topological Lemma*). Let $\{f_\alpha\}_{\alpha \in A}$ *be a family of real-valued functions in X. Let*

$$f_A(x) = \inf_{\alpha \in A} f_\alpha(x).$$

Then there is a countable subset A_0 of A such that

$$(f_{A_0})^\wedge = (f_A)^\wedge.$$

We may, without loss of generality, suppose $f_\alpha(X) \subset [-1, 1]$ for all $\alpha \in A$. Let $\{G_n\}$ be a base for X such that each G_n appears in the sequence infinitely often. For example, the following arrangement ensures this:

$$G_1, G_2, G_1, G_2, G_3, G_1, G_2, G_3, G_4 \ldots$$

For each n choose $\alpha_n \in A$ so that

$$\inf_{y \in G_n} f_{\alpha_n}(y) - \inf_{y \in G_n} f_A(y) < \frac{1}{n} \tag{4.8}$$

Let $A_0 = \{\alpha_n\}$. Suppose that g is l.s.c. in X and $g \leqslant f_{\alpha_n}$ for all n. Given $a \in X$ and $\varepsilon > 0$, there is a neighbourhood N of a such that, when $x \in N$, $g(x) > g(a) - \frac{1}{2}\varepsilon$, and then there is a G_k containing a in which this inequality holds. Furthermore, we may choose k so that $1/k < \frac{1}{2}\varepsilon$.

Consequently,

$$g(a) - \inf_{y \in G_k} g(y) < \tfrac{1}{2}\varepsilon, \tag{4.9}$$

and, furthermore,

$$\inf_{y \in G_k} g(y) - \inf_{y \in G_k} f_{\alpha_k}(y) \leqslant 0. \tag{4.10}$$

By (4.8), (4.9) and (4.10)

$$g(a) - \inf_{y \in G_k} f_A(y) < \varepsilon$$

and hence $g(a) - (f_A)^\wedge(a) < \varepsilon$.

Since ε is arbitrary this gives $g(a) \leqslant (f_A)^\wedge(a)$. Now $(f_{A_0})^\wedge$ constitutes a particular case of such a g, and so $(f_{A_0})^\wedge$ is not greater than $(f_A)^\wedge$. But, since $A_0 \subset A$, we also have the converse inequality; and the theorem now follows.

THEOREM 4.34. *Suppose that $\{v_n\}$ is a decreasing sequence of functions* *each of which is superharmonic and non-negative in X. Let* $v = \lim_{n\to\infty} v_n$. *Then* $\hat{v}$ *is superharmonic and*

$$\hat{v} = v \text{ quasi-everywhere.}$$

First, for each regular set B,

$$v(a) = \lim_{n\to\infty} v_n(a) \geqslant \lim_{n\to\infty} \int v_n(\xi)\, \sigma_a^B(d\xi) = \int v(\xi)\, \sigma_a^B(d\xi)$$

and then, since the last term is continuous,

$$\hat{v}(a) \geqslant \int v(\xi)\, \sigma_a^B(d\xi).$$

Since $v \geqslant \hat{v}$, we have $\hat{v}(a) \geqslant \int \hat{v}(\xi)\, \sigma_a^B(d\xi)$ and so, $\hat{v}$ being l.s.c. in X, is superharmonic there.

Now suppose, for the time being, that $\{v_n\}$ is locally bounded. Then we must show that the set $E = \{x \mid v(x) > \hat{v}(x)\}$ is polar. Now $E = \bigcup_{k=1}^{\infty} E_k$ where $E_k = \{x \mid v(x) - \hat{v}(x) > 1/k\}$. It is therefore enough to show that each E_k is polar, and so enough to show that, given $\varepsilon > 0$, the set $A = \{x \mid v(x) - \hat{v}(x) > \varepsilon\}$ is polar. Now

$$A = \bigcap_{n=1}^{\infty} \{x \mid v_n(x) - \hat{v}(x) > \varepsilon\} = \bigcap_{n=1}^{\infty} A_n, \text{ say,}$$

and

$$A_n = \bigcap_{m=1}^{\infty} \{x \mid v_n(x) - f_m(x) > \varepsilon\} = \bigcap_{m=1}^{\infty} A_{mn}, \text{ say,}$$

where $\{f_m\}_{m=1,2,\ldots}$ is an increasing sequence of continuous functions converging to $\hat{v}$. Each A_{mn} is open, so each A_n is a G_δ and so A is a G_δ. Since X has a countable base, any open set in X is an F_σ† and thus A is an $F_{\sigma\delta}$ which is a $K_{\sigma\delta}$.

To prove A polar, it is enough to show that, when p is a continuous potential, $*p^A(a) = 0$ at some point (Theorem 4.29, in the course of the proof). Now $*p^A(a)$ is a generalised capacity, by Theorem 4.10, so A being a $K_{\sigma\delta}$, is capacitable (Theorem 3.42) and hence

$$*p^A(a) = \sup\{*p^K(a) \mid K \subset A, K \text{ compact}\}.$$

It is therefore enough to show that, for every compact $K \subset A$, we have

† Since X is locally compact and has a countable base it is metrisable (Bourbaki, Livre III, Chapitre 9,43) and, if d is a metric in X and $G \subset X$ is open

$$G = \bigcup_{n=1}^{\infty} \left\{x \in X \mid \text{dist}(x, \complement G) \geqq \frac{1}{n}\right\}.$$

${}^*p^K(a) = 0$. Given such a K, choose $K \subset G \subset X$ open and relatively compact. Then, for $x \in G \setminus K$, and for each n,

$$v_n(x) \geqslant H_{v_n}^{G\setminus K}(x) \geqslant H_v^{G\setminus K}(x)$$

(both solution functions exist, since v_n is l.s.c. and v is bounded Borel). Hence $v(x) \geqslant H_v^{G\setminus K}(x)$ and, since the right-hand side is continuous, $\hat{v}(x) \geqslant H_v^{G\setminus K}(x) \geqslant H_{\hat{v}}^{G\setminus K}(x)$.

By Theorem 4.32 $H_v^{G\setminus K}$ is the greatest harmonic minorant of $\hat{v}$ in $G\setminus K$; and so $H_v^{G\setminus K}(x) = H_{\hat{v}}^{G\setminus K}(x)$ and hence $H_{v-\hat{v}}^{G\setminus K}(x) = 0$.

Now $v(x) - \hat{v}(x) > \varepsilon$ for $x \in K$. Let p be a continuous potential in X, and choose $\lambda > 0$ so that $\lambda p(x) < \varepsilon$ for $x \in K$. Define f on $\partial(G\setminus K) = \partial G \cup \partial K$ by

$$f(\xi) \begin{cases} = 0, & \xi \in \partial G \\ = \lambda p(\xi), & \xi \in \partial K \end{cases}$$

Then ${}^*(\lambda p)_G^K = H_f^{G\setminus K} \leqslant H_{v-\hat{v}}^{G\setminus K}$ in $G\setminus K$ (${}^*(\lambda p)_G^K$ is the pre-sweep of λp regarding G as the total space) and thus ${}^*p_G^K(x) = 0$ for $x \in G\setminus K$. Thus, for some $a \in X$, ${}^*p_G^K(a) = 0$, and so K is polar in G. Since G is arbitrary and X has a countable base, K is polar (in X).

This, then, shows that $\hat{v} = v$ quasi-everywhere if $\{v_n\}$ is locally bounded. If $\{v_n\}$ is not locally bounded, let

$$v_n^k = \min\,[v_n, kp\}.$$

Then $\{v_n^k\}$ decreases with n and $w_k = \lim_{n\to\infty} v_n^k = \min\,[v, kp]$. Furthermore,† $\hat{w}_k = \min\,[\hat{v}, kp]$ and so $\lim_{k\to\infty} \hat{w}_k = \hat{v}$, and $\lim_{k\to\infty} w_k = v$.

Since $\{v_n^k\}$ is, for each k, a locally bounded decreasing sequence of non-negative superharmonic functions, $w_k = \hat{w}_k$ quasi-everywhere; and so $\{w_k(x)\} \neq \{\hat{w}_k(x)\}$ in, at most, a countable union of polar sets, which is again polar. Hence

$$\hat{v} = \lim_{k\to\infty} \hat{w}_k = \lim_{k\to\infty} w_k = v \text{ quasi-everywhere.}$$

COROLLARY 4.34.1. *Let* $\{v_n\}$, $v_n \geqslant 0$, *be a sequence of hyperharmonic functions. Then* $\inf_n v_n$ *and* $\liminf_{n\to\infty} v_n$ *are either quasi-superharmonic (that is, equal to a superharmonic function quasi-everywhere) or are everywhere equal to* $+\infty$.

† We have, quite generally,

$$\liminf_{x\to a} \min\,[f(x), g(x)] = \min\,[\liminf_{x\to a} f(x), \liminf_{x\to a} g(x)].$$

Let $w_n^k = \min_{n \leqslant r \leqslant n+k} v_r$. Then either w_n^k is superharmonic or each of v_r, $n \leqslant r \leqslant n+k$ is $+\infty$. Also

$$\lim_{k \to \infty} w_1^k = \inf_n v_n.$$

If any v_n is not $+\infty$, then the tail of $\{w_1^k\}$ is a decreasing sequence of superharmonic functions, and so $\inf_n v_n$ is quasi-superharmonic. Otherwise $\inf_n v_n = +\infty$, and then $\liminf_{n \to \infty} v_n = +\infty$.

Similarly, $w_n = \lim_{k \to \infty} w_n^k = \inf_{r \geqslant n} v_r$ is, if any v_n is not $+\infty$, quasi-superharmonic. So $w_n = \hat{w}_n$ quasi-everywhere. Since $\hat{w}_n$ is increasing and superharmonic, $\lim \hat{w}_n$ is either superharmonic or $+\infty$ and thus $\liminf_{n \to \infty} v_n = \lim_{n \to \infty} w_n$ is quasi-superharmonic or identically $+\infty$.

THEOREM 4.35. *Let $\{v_\alpha\}_{\alpha \in A}$ be a family of superharmonic functions each $\geqslant 0$. Then, if $v_A = \inf_A v_\alpha$, we have $v_A = (v_A)^\wedge$ quasi-everywhere and $(v_A)^\wedge$ is superharmonic.*

By Theorem 4.33 there is a countable subset A_0 of A such that $(v_{A_0})^\wedge = (v_A)^\wedge$ and $(v_A)^\wedge \leqslant v_A \leqslant v_{A_0}$. By Corollary 4.34.1 $(v_{A_0})^\wedge = v_A$ quasi-everywhere, and so the result follows.

This last theorem makes precise a remark in Chapter 2 about down-directed sets of hyperharmonic functions, and shows indeed that, as Theorem 4.4(2) suggests, it is the infimum of any family rather than that of a down-directed family that is quasi-superharmonic.

LEMMA 4.36. *Let u be a positive superharmonic function in X. Then, for any set E,*

(1) $u^E(x) = {}^*u^E(x)$ *quasi-everywhere.*

(2) $u^E(x) = {}^*u^E(x)$ *for $x \in X \setminus E$.*

Since the pre-sweep is the infimum of a family of positive superharmonic functions, Theorem 4.35 applies; and (1) follows.

Let w be associated with the polar set in which the equality does not hold. Then, for $\lambda > 0$, $u^E + \lambda w \geqslant {}^*u^E$ everywhere in E and so, w being superharmonic and positive, everywhere. Given $a \in X \setminus E$, w can have been so chosen that $w(a) < +\infty$. (For let $\{G_n\}$ be a sequence of open sets with $\bigcap_{n=1}^{\infty} G_n = (a)$. Then $E \setminus G_n$ is polar and we may associate $w_n \geqslant 0$ with $E \setminus G_n$ such that $w_n(a) < 1/n^2$. Then $w = \sum_{n-1}^{\infty} w_n$ takes the value $+\infty$ on E and $w(a) < +\infty$). Then, since the inequality holds for every $\lambda > 0$, $u^E(a) \geqslant {}^*u^E(a)$; which gives (2).

THEOREM 4.37. *The set of points at which a set is thin is polar.*

Let p be a continuous potential in X, and let $\{G_k\}$ be a countable base for X. If E is thin at $a \in E$ then $\{a\}$ is polar and $E \setminus \{a\}$ is thin at a. We can, by Theorem 4.30, find G_k containing a such that

$$p^{E \cap G_k}(a) = p^{E \cap G_k \setminus \{a\}}(a) \leqslant {}^*p^{E \cap G_k \setminus \{a\}}(a) < p(a).$$

Hence $a \in \{x \in E \cap G_k \mid p^{E \cap G_k}(x) < p(x)\}$. By Lemma 4.36 and Theorem 4.9(2), $p^{E \cap G_k}(x) = p(x)$ quasi-everywhere in $E \cap G_k$. So this last set is polar, and, since $E = \bigcup_{k=1}^{\infty} E \cap G_k$, the set of points in E at which E is thin is polar.

We have already seen in Theorem 4.31 that, if $X \setminus G$ is thin at $a \in \partial G$, then a is irregular for G. We may now show that

a is regular for G if, and only if, $X \setminus G$ is not thin at a.

For, first, if $\{a\}$ is not polar, then $F = X \setminus G$ is not thin at a. Let p be a continuous potential, and let N be a neighbourhood containing a. Then, by Lemma 4.36(1), since $\{a\}$ is not polar,

$$^*p^{F \cap N}(a) = p^{F \cap N}(a).$$

Since $a \in F \cap N$, $^*p^{F \cap N}(a) = p(a)$, by Theorem 4.9(2); and so $p^{F \cap N}(a) = p(a)$. Hence, by Theorem 4.27, a is regular for G.

If $\{a\}$ is polar, then $p^{F \cap N}(a) = p^{F \cap N \setminus \{a\}}(a)$ and, by Lemma 4.36(2), $p^{F \cap N \setminus \{a\}}(a) = {}^*p^{F \cap N \setminus \{a\}}(a)$, so that $p^{F \cap N}(a) = {}^*p^{F \cap N \setminus \{a\}}(a)$. Thus we have

$$p^{F \cap N}(a) = p(a) \tag{4.11}$$

if, and only if,

$$^*p^{F \cap N \setminus \{a\}}(a) = p(a). \tag{4.12}$$

Now (Theorem 4.27), (4.11) is the condition that a be regular for G and (Theorem 4.30), (4.12) is the condition that F be not-thin at a. This gives the result.

We are now, after many tribulations, in a position to bring the matter of the Dirichlet Problem to a satisfying conclusion.

THEOREM 4.38. *Let X be a locally compact connected Hausdorff space with a countable base. Let Axioms* 1, 2, 3, *P and D hold in X. Then, given a relatively compact open set G, and given $f \in \mathscr{C}(\partial G)$, there is a unique bounded $h \in \mathscr{H}_G$ such that*

$$\lim_{x \in G,\, x \to \xi} h(x) = f(\xi)$$

at every regular point of ∂G which is to say, at every point of ∂G, except possibly those of a polar set.

First, the set of points irregular for G is the set at which $X \setminus G$ is thin; and this, by Theorem 4.37, is a polar set. Second, the generalised solution H_f^G exists and tends to f at every regular point. It therefore remains to show that h is unique.

Let, then, $h \in \mathscr{H}_G$ be bounded and such that $\lim_{x \in G, x \to \xi} h(x) = 0$ at every point regular for G. Define g in ∂G by

$$g(\xi) = \limsup_{x \in G, x \to \xi} h(x).$$

Let w be associated with the set E of irregular points for G and, given $a \in G$, suppose $w(a) < +\infty$. Then, for each n,

$$\liminf_{x \to \xi} \frac{1}{n} w(x) \geqslant 0 \text{ for } \xi \in \partial G; \liminf_{x \to \xi} \frac{1}{n} w(x) = +\infty \text{ for } \xi \in E.$$

Consequently, for $\xi \in \partial G$,

$$\liminf_{x \in G, x \to \xi} \left(\frac{1}{n} w(x) - h(x) \right) = \liminf_{x \in G, x \to \xi} \left(\frac{1}{n} w(x) \right) - \limsup_{x \in G, x \to \xi} h(x) \geqslant 0$$

and so $\frac{1}{n} w(x) \geqslant h(x)$ for $x \in G$, and hence $h(a) \leqslant \frac{1}{n} w(a)$ for every n and so $h(a) = 0$. Arguing in a similar way with $-h$ we find $h(a) \leqslant 0$ and hence $h(a) = 0$. Thus $h = 0$ in G.

Finally, suppose that h is a quasi-solution to the Dirichlet Problem, that is, that

$$\lim_{x \in G, x \to \xi} h(x) = f(\xi)$$

at every regular point. Then $h(x) - H_f^G(x) \to 0$ at every regular point of ∂G, and so $h(x) = H_f^G(x)$, giving the uniqueness of h.

§ 4.7. The Laplace Case for $\mathbb{R}^p$

We turn now to the case of the Laplace equation thus returning to the classical harmonic functions which served as guide and model for the abstract theory.

There is a sharp distinction between the cases $p = 2$ and $p \geqslant 3$. When $p = 2$ the theory fits the space $B_1(0)$ rather than $\mathbb{R}^p$, whereas when $p \geqslant 3$ it applies directly to $\mathbb{R}^p$. So we shall suppose that the space Y is given by $B_1(0)$ when $p = 2$ and by $\mathbb{R}^p$ when $p \geqslant 3$.

In any open set $G \subset Y$ the solutions of the Laplace equation

$$\frac{\partial^2 V}{\partial x_1^2} + \ldots + \frac{\partial^2 V}{\partial x_p^2} = 0$$

clearly constitute an $\mathscr{H}_G$, and so Axiom 1 is satisfied.

The open balls $B_r(a)$ are regular sets since, by Theorem 2.17, the Poisson integral furnishes the required unique solution to the Dirichlet Problem. Also when $f \geqslant 0$, $I_r^a f \geqslant 0$. Since the open balls form a neighbourhood base for Y, Axiom 2 is satisfied for Y.

Theorem 2.24 shows that Axiom 3 is satisfied for Y.

When $p \geqslant 3$, $|x|^{2-p}$ is superharmonic and positive in $\mathbb{R}^p$ and is not harmonic there. When $p = 2$, $\log(1/|x|)$ is positive, superharmonic and not harmonic in $B_1(0)$. Thus Axiom P is satisfied for Y.

It remains to show, and it is by no means a trivial matter, that Axiom D is satisfied in Y. Before we can prove the Domination Principle itself, we need a preliminary lemma.

LEMMA 4.39. *A necessary and sufficient condition that a function f positive and superharmonic in Y be a Newtonian potential is that $\mathscr{S}_r^0(f) \to 0$ as $r \to 1$ when $p = 2$ and as $r \to \infty$ when $p \geqslant 3$.*

We shall give the proof in detail for $p \geqslant 3$ and indicate the slight changes needed for $p = 2$.

Suppose first that U^μ is a Newtonian potential which is not identically $+\infty$. Then U^μ is superharmonic, so locally integrable and consequently $\mathscr{S}_r^0(U^\mu) < \infty$. Also

$$\mathscr{S}_r^0(U^\mu) = \int U^{\sigma_r}(\xi)\,\mu(d\xi)$$

where σ_r denotes σ_r^0 and, by Lemma 2.15, $U^{\sigma_r}(\xi) \leqslant h(r)$ and so decreases to 0 as $r \to \infty$. Consequently, by Theorem 1.22, $\mathscr{S}_r^0(U^\mu)$ tends to 0 as r tends to ∞. This proves the necessity.

Conversely, suppose that f is positive and superharmonic in Y and such that

$$\lim_{r\to\infty} \mathscr{S}_r^0(f) = 0. \tag{4.13}$$

Let $\mu = C\Delta f$, where Δ is to be understood in the generalised sense of § 6 of Chapter 2. Then, for $0 < r < s < \infty$ we have, by the Riesz Decomposition (Theorem 2.41)

$$f(x) \begin{cases} = U^{\mu_r}(x)+g_r(x), & x \in B_r(0), \\ = U^{\mu_s}(x)+g_s(x), & x \in B_s(0), \end{cases}$$

where μ_r is the restriction of μ to $B_r(0)$, and g_r is harmonic. Thus

$$f(x)-U^{\mu_r}(x) = U^{\mu_s-\mu_r}(x)+g_s(x), \quad x \in B_s(0),$$

and so $f-U^{\mu_r}$ is superharmonic in $B_s(0)$ for all $s \geqslant r$ and so $f-U^{\mu_r}$ is superharmonic in Y. Also, for $x \in B_r(0)$, $U^{\mu_s-\mu_r}(x) = g_r(x)-g_s(x)$ so that $f-U^{\mu_r}$ is harmonic in $B_r(0)$.

Now $\mathcal{S}_\rho^0(U^{\mu_r}) \to 0$ as $\rho \to \infty$, and this, with (4.13), shows that $\lim_{\rho\to\infty} \mathcal{S}_\rho^0(f-U^{\mu_r}) = 0$. Furthermore, by Theorem 2.30, this mean decreases with increasing ρ and so $\mathcal{S}_\rho^0(f-U^{\mu_r}) \geqslant 0$ for all ρ. The same considerations apply to $\mathcal{S}_\rho^a(f-U^{\mu_r})$ and so

$$f(a)-U^{\mu_r}(a) \geqslant \mathcal{S}_\rho^a(f-U^{\mu_r}) \geqslant 0.$$

Thus $f-U^{\mu_r} \geqslant 0$ everywhere in Y. Also $U^\mu(x) = \lim_{r\to\infty} U^{\mu_r}(x) \leqslant f(x)$ for $x \in Y$, and so U^μ is superharmonic in Y. Then $\mathcal{S}_\rho^0(f-U^\mu)$ is not greater than $\mathcal{S}_\rho^0(f-U^{\mu_r})$ and so $\lim_{\rho\to\infty} \mathcal{S}_\rho^0(f-U^\mu) = 0$. Further, $f-U^\mu = \inf_r (f-U^{\mu_r})$ and so $f-U^\mu$ is harmonic in $B_r(0)$ for every r and hence in Y.

Finally, given $\varepsilon > 0$, we can find ρ such that $\mathcal{S}_\rho^0(f-U^\mu) < \varepsilon$ and so $(f-U^\mu)(0) < \varepsilon$. Since ε is arbitrary, this gives $(f-U^\mu)(0) = 0$. Since $f-U^\mu$ is non-negative and harmonic, it is thus identically 0 and so $f(x) = U^\mu(x)$. This gives the lemma when $p \geqslant 3$.

When $p = 2$ the above proof applies if we substitute '$\to 1$' for '$\to \infty$' wherever it appears.

A Newtonian potential in Y is a potential in the sense of the general theory, since its greatest harmonic minorant is zero. For, if u is harmonic and $0 \leqslant u \leqslant U^\mu$ in Y then $\mathcal{S}_r^0(u) \to 0$ as $r \to \infty$ ($r \to 1$) and so, by Theorem 2.46 or, more simply, by the above argument, $u = 0$.

THEOREM 4.40 (*The Domination Principle*). *Suppose that p is a locally bounded potential in Y, which is harmonic precisely in the open set G, and that v is positive and superharmonic in Y and such that $v(x) \geqslant p(x)$ for $x \in Y\setminus G$. Then $v(x) \geqslant p(x)$ everywhere in Y.*

Again, we shall deal in detail with the case $p \geqslant 3$.

Let $p = U^\mu$ and let μ_n be the restriction of μ to $B_n = B_n(0)$. Then

$$\| \mu_n \|^2 = \int U^{\mu_n}(x)\, \mu_n(dx) \leqslant \sup_{B_n} p\, \mu(B_n)$$

and so μ_n is of finite energy. Now

$$\mathcal{S}_r^0\, (\inf\,[U^{\mu_n}, v]) \leqslant \mathcal{S}_r^0(U^{\mu_n})$$

and $\mathcal{S}_r^0(U^{\mu_n})$ converges to 0 as $r \to \infty$. So therefore does the left-hand side and is therefore, by Lemma 4.39, a potential since $\inf\,[U^{\mu_n}, v]$ is positive and superharmonic. Set $U^\nu = \inf\,[U^{\mu_n}, v]$.

Since $U^\nu \leqslant U^{\mu_n}$, ν is of finite energy and

$$\| \mu_n - \nu \|^2 = \int (U^{\mu_n} - U^\nu)\, \mu_n(dx) - \int (U^{\mu_n} - U^\nu)\, \nu(dx).$$

Now $U^\nu = U^{\mu_n}$ in supp μ_n, since supp $\mu_n \subset$ supp $\mu = X\setminus G$, and $U^\nu \leqslant U^{\mu_n}$ everywhere; so the first integral vanishes and the second is non-negative.

Consequently $\| \mu_n - \nu \| = 0$ and thus, by Lemma 3.6, $\mu_n = \nu$. Hence $U^{\mu_n} = U^{\nu}$ and so $U^{\mu_n} \leqslant v$ everywhere. Thus

$$p = U^{\mu} = \lim_{n \to \infty} U^{\mu_n} \leqslant v \text{ everywhere in } Y$$

and this gives the domination principle.

When $p = 2$, the proof above applies if we substitute '$\to 1$' for '$\to \infty$' and '$B_{1-(1/n)}(0)$' for '$B_n(0)$'.

This then shows that Axiom D holds for Y. Thus Y is a space to which all the conclusions of the general theory apply. In particular, therefore, the Dirichlet Problem can be solved in the sense of Theorem 4.38 for every bounded open set in $\mathbb{R}^p$. When $p = 2$, a suitable uniform contraction in $\mathbb{R}^p$ will take any bounded open set into one which is relatively compact in $B_1(0)$. For this set the Dirichlet Problem can be solved. Since now the Laplace equation remains invariant under such a transformation, the Dirichlet Problem is solved for any bounded open set.

It is of interest to see what 'polar' and 'thin' mean in this context. The next lemma shows the connection between polar and outer 2-capacity when $p \geqslant 3$.

LEMMA 4.41. *A set $E \subset \mathbb{R}^p$, $p \geqslant 3$, is polar if, and only if, its outer 2-capacity is zero.*

Suppose that E is polar and bounded. Then we can find a function f superharmonic in $\mathbb{R}^p$ and taking the value $+\infty$ in E. Now, given any bounded open set $G \supset E$ and setting $\mu_G = \operatorname{rest}_G (C\Delta f)$ we have, for $x \in G$,

$$f = U^{\mu_G} + h, \qquad h \text{ harmonic},$$

and so $U^{\mu_G}(x) = +\infty$ for $x \in E$.

Given $\lambda > 0$, $G_\lambda = \{x \in G \mid U^{\mu_G}(x) > \lambda\}$ is open. Suppose that the compact K is contained in G_λ and that γ_K is the equilibrium measure for K. Then

$$\operatorname{cap} K = \int \gamma_K(dt) \leqslant \frac{1}{\lambda} \int U^{\mu_G}(t)\, \gamma_K(dt) = \frac{1}{\lambda} \int_G U^{\gamma_K}(t) \mu(dt) \leqslant \frac{1}{\lambda}\, \mu(G).$$

Hence $\operatorname{cap}_* G_\lambda = \operatorname{cap}^* G_\lambda \leqslant \mu(G)/\lambda$ so $\operatorname{cap}^* E \leqslant \mu(G)/\lambda$ for every $\lambda > 0$. Thus $\operatorname{cap}^* E = 0$.

If E is not bounded, let $E_n = E \cap B_n(0)$. Then E_n is polar, so $\operatorname{cap}^* E_n = 0$. Since $\bigcup_{n=1}^{\infty} E_n = E$, $\operatorname{cap}^* E = \lim \operatorname{cap}^* E_n = 0$.

Conversely, suppose $\operatorname{cap}^* E = 0$, and suppose for the moment that E is bounded. Choose a so that dist $(a, E) > 2$. For every n we can find an open set G_n containing E such that dist $(a, G_n) > 1$ and cap $G_n < n^{-2}$. Furthermore, there is an increasing sequence of compact sets $\{K_{np}\}$ with $K_{np} \subset \mathring{K}_{np+1}$ such that $G_n = \bigcup_{p=1}^{\infty} K_{np}$.

Now $U^{\gamma K_{np}}(x) = 1$ for $x \in \mathring{K}_{np}$ and $U^{\gamma K_{np}}$ is superharmonic. Also $v_n(x) = \lim_{p \to \infty} U^{\gamma K_{np}}(x)$ takes the value 1 on G_n and is superharmonic. Further

$$U^{\gamma K_{np}}(a) = \int |x-a|^{2-p} \gamma_{K_{np}}(dx) \leqslant \text{cap } K_{np} \leqslant \text{cap } G_n < n^{-2},$$

and so $v_n(a) < n^{-2}$. Consequently, if $w = \sum_{n=1}^{\infty} v_n$, then $w(a)$ is finite, and so w is superharmonic. Also $w(x) \geqslant \sum_{r=1}^{n} v_r(x)$ and this last takes the value n in G_n. Hence $w(x) = +\infty$ for $x \in E$. So E is polar.

If E is not bounded, let $E_n = E \cap B_n(0)$. Then cap * $E_n = 0$ and E_n is bounded, so E_n is polar. Consequently $E = \bigcup_{n=1}^{\infty} E_n$ is polar.

We remark that if $p = 2$ the result corresponding to Lemma 4.41 involves logarithmic capacity in place of Newtonian capacity.

Lemma 4.41 enables us to say that a set E is polar only if there is no non-zero measure μ supported on E for which $U^{\mu}(x)$ remains bounded for all $x \in \mathbb{R}^p$.

It is a relatively simple matter to show, by using Lebesgue q-dimensional measure, that, if L is a q-dimensional linear variety in $\mathbb{R}^p$, then L is of positive α-capacity for every $\alpha < q$ and that L is of zero q-capacity. Reference may be made to du Plessis [1], [2] for details. In fact Lemma 5 in [2] shows that any bounded subset of L is of zero outer q-capacity, but since any subset of L is the union of a countable family of bounded subsets, the result follows readily for the variety L itself. This means, for example that a plane in $\mathbb{R}^3$ is not polar while a line in $\mathbb{R}^3$ is.

We turn now to thinness. In $\mathbb{R}^p$ this has a marked geometrical connotation. The Lebesgue potential in $\mathbb{R}^p$ is given, for $p \geqslant 3$, by

$$V(x) = \int_0^1 t^{p-2} [(x_p - t_p)^2 + x_2^2 + \ldots + x_{p-1}^2]^{1-\frac{1}{2}p} \, dt.$$

We then have $V(0) = 1$, and $V(x)$ tends to $V(0)$ as x tends to 0 along any fixed direction other than the positive x_1-axis. Furthermore, for $p \geqslant 4$, we have $V(x) > \lambda > 1$ when x lies in a sharp spine given by

$$\sqrt{(x_1^2 + \ldots + x_{p-1}^2)} < \left(\frac{K}{\lambda}\right)^{\frac{1}{p-3}} x_p^{\frac{p-2}{p-3}}.$$

When $p = 3$, the region for which $V(x) > \lambda > 1$ is exponentially sharp. It is given by

$$\sqrt{(x_1^2+x_2^2)} < K \exp(-\phi(x_3)/x_3)$$

where $\lim_{x_3 \to 0} \phi(x_3) = \frac{1}{2}(\lambda-1)$. For the case $p \geqslant 4$ see du Plessis [1] and for $p = 3$ see Brelot [1], p. 79.

Thus, if we set $G = B_1(0) \setminus \{x \in B_1(0) \mid V(x) > \lambda\}$ then the point $0 \in \partial G$ is irregular for G, since V is a superharmonic function such that

$$\liminf_{x \in \complement G, x \to 0} V(x) > V(0).$$

This therefore suggests that one could picture an irregular point as being the apex of a sharp spike which has been, as it were, driven into the set G from outside. In this connection, see an interesting intuitive description in Kellogg, p. 285. This is, inevitably, much too simple to take account of the real pathology of these points but it does, to some extent, reflect the truth of the situation. This is shown by the form which an important criterion for irregularity, due to Wiener, takes.

Wiener's Criterion states that:

A point $a \in \partial G$ is irregular for G if, and only if, for any $k > 1$, and for

$$S_n = \{x \in \mathbb{R}^p \mid k^n \leqslant h_a(x) < k^{n+1}\}$$

we have

$$\sum k^n \operatorname{cap}(\complement G \cap S_n) < +\infty.$$

The reader is referred to Brelot [5] for a discussion.

We may use this criterion to show that any conical spike driven into a 'solid' G does not have at its vertex an irregular point. For if $S = \{x \mid \sqrt{(x_1^2+\ldots+x_{p-1}^2)} \leqslant Kx_p\}$ is such a conical spike and $G = \{x \mid 0 \leqslant x_i \leqslant k^{1/(2-p)}\} \setminus S$ then $[G \cap S_n = S_n = k^{n/(2-p)} S_1$; so, by Lemma 3.30(2), $\operatorname{cap} S_n = k^{-n} \operatorname{cap} S_1$ and, since $\mathring{S}_1 \neq \phi$, $\operatorname{cap} S_1 > 0$. Hence the series in the Wiener criterion becomes $\Sigma \operatorname{cap} S_1$, which is divergent.

In particular, this effectively shows that any point of ∂G at which an exterior sphere condition operates must also be a regular point.

We have seen that the solution of the Dirichlet Problem matches the boundary data quasi-everywhere, and we can show that this, in a sense, is a best possible result. One can produce a region G with a connected boundary, on which the set of irregular points is of positive α-capacity for every $\alpha < (p-2)$. There is in du Plessis [1] a very complicated such example for $p \geqslant 4$ and also a very simple one for $p \geqslant 3$ in du Plessis [2]. This latter gives, in $\mathbb{R}^3$, a square candle in which that part of the wick within the candle, regarded as part of the boundary of the region is wholly composed of irregular points. Being a line, the wick is polar but also of positive α-capacity for $\alpha < 1$.

On the other hand, it is wrong to suppose that, if all the points of the boundary are regular, then the boundary is smooth. An example in du Plessis [2] of a domain in $\mathbb{R}^p$ ($p \geqslant 2$) with a connected frontier of **positive** Lebesgue p-measure of which every point is regular serves to show just how rough the boundary can be even when the domain is well behaved from the Dirichlet point of view!

We turn next to the question of the Green's function. Suppose we place a unit positive charge at a point inside a conductor. Then, according to electrostatic lore, it will induce a positive charge on the outer surface of the conductor, and the potential of this latter charge will coincide with the potential of the original point charge everywhere outside the conductor. The difference between the two potentials is zero outside the conductor, and is harmonic inside the conductor except at the seat of the point charge, in the neighbourhood of which it approximates $1/r$, where r is the distance from the point charge, and is zero on the conductor. Thus, on physical grounds, we might, with Green, assert the existence within the conductor of a function of the form

$$1/r + \text{harmonic function}$$

which vanishes on the conductor, that is, of what is now called Green's function.

To turn this into mathematics we proceed as follows:

Suppose that G is a bounded open set and that $a \in \mathbb{R}^p$. Define f in ∂G by

$$f(\xi) = -h_a(\xi).$$

Then the **Green function with pole a for G,** $g_a^G(x)$ is defined, for $x \in G$, by

$$g_a(x) = h_a(x) + H_f^G(x).$$

We will denote it simply by $g_a(x)$ when there can be no ambiguity about G. If $a \notin G$, then $h_a(x)$ is harmonic in G and $\lim_{x \to \xi} h_a(x) = h_a(\xi)$ when $\xi \in \partial G$. Hence $H_f^G(x) = -h_a(x)$ and so $g_a(x) = 0$ identically in G.

When $a \in G$ we have, by Theorem 4.38,

$$\lim_{x \in G, x \to \xi} g_a(x) = 0 \text{ quasi-everywhere in } \partial G.$$

In fact, in every component of G other than the one containing a, $g_a(x)$ is identically zero. **In the component containing a, $g_a(x) > 0$.** For let $w \in \mathcal{O}_f^G$. Then

$$\liminf_{x \in G, x \to \xi} (h_a(x) + w(x)) = h_a(\xi) + \liminf_{x \in G, x \to \xi} w(x) \geqslant h_a(\xi) - h_a(\xi) = 0.$$

Since $h_a+w \in \mathscr{H}_G^*$, we have $h_a(x)+w(x) \geqslant 0$ for $x \in G$. Since this holds for all such w, this gives $h_a(x)+H_f^G(x) \geqslant 0$ for $x \in G$. Now $h_a+H_f^G \in \mathscr{H}_G^*$ and is not identically zero, since $h_a(a)+H_f^G(a) = +\infty$. Hence $g_a(x) = h_a(x)+H_f^G(x) > 0$ for $x \in G$.

Consider the set

$$\mathscr{G}_a = \{h_a+u \mid u \in \mathscr{H}_G \text{ and } h_a(x)+u(x) \geqslant 0,\ x \in G\}.$$

Since, for any such u, $u(x) \geqslant -h_a(x)$, and so $\liminf_{x\in G, x\to\xi} u(x) \geqslant -h_a(\xi)$ for $\xi \in \partial G$, $u \in \mathscr{O}_f^G$, and so $u(x) \geqslant H_f^G(x)$ for $x \in G$. Also $g_a \in \mathscr{G}_a$. Thus g_a^G *is the smallest member of* $\mathscr{G}_a$. This property is useful in the sequel.

THEOREM 4.42.

(1) *Suppose* $G_1 \subset G_2$. *Then* $g_a^{G_1} \leqslant g_a^{G_2}$.

(2) *Suppose that* $\{G_n\}$ *is an increasing sequence of open sets, and that* $G = \bigcup_{n=1}^{\infty} G_n$ *is bounded. Then* $g_a^{G_n}(x)$ *converges increasingly to* $g_a^G(x)$.

(3) *Suppose that* $x, y \in G$. *Then* $g_x^G(y) = g_y^G(x)$.

Let $w \in \mathscr{O}_{-h_a}^{G_2}$. Then, since w is superharmonic and $\partial G_1 \subset \bar{G}_2$, we have, when $\xi \in \partial G_1 \cap G_2$, $\liminf_{x\in G_1, x\to\xi} w(x) \geqslant \liminf_{x\to\xi} w(x) = w(\xi) \geqslant H_{-h_a}^{G_2}(\xi) \geqslant -h_a(\xi)$ and when $\xi \in \partial G_1 \cap \partial G_2$, $\liminf_{x\in G_1, x\to\xi} w(x) \geqslant \liminf_{x\to\xi} w(x) \geqslant -h_a(\xi)$ and so $w \in \mathscr{O}_{-h_a}^{G_1}$. Thus $\mathscr{O}_{-h_a}^{G_2} \subset \mathscr{O}_{-h_a}^{G_1}$, and so

$$H_{h_a}^{G} = \inf\{w \mid w \in \mathscr{O}_{h_a}^{G_1}\} \leqslant \inf\{w \mid w \in \mathscr{O}_{h_a}^{G_2}\} = H_{h_a}^{G_2}$$

and hence $g_a^{G_1} \leqslant g_a^{G_2}$, giving (1).

Let $g_n = g_a^{G_n}$, $g = g_a^G$. Then, by (1), $g_n \leqslant g$, and so $\lim_{n\to\infty} g_n \leqslant g$. Let $g_n = h_a+u_n$. Now $g_n \geqslant 0$ in G_n, and so $\lim_{n\to\infty} g_n \geqslant 0$ in G.

Furthermore $\{u_n\}$ is increasing and bounded above, and u_n is harmonic, so $u = \lim_{n\to\infty} u_n$ is harmonic in each G_n and hence in G. Thus $\lim_{n\to\infty} g_n = h_a+u \in \mathscr{G}_a$, and so $\lim_{n\to\infty} g_n \geqslant g$, since g is the least member of $\mathscr{G}_a$. This gives (2).

We have $g_y^G(x) = g_y(x) = h_y(x) - \int h_y(\xi)\, \sigma_x^G(d\xi)$

$$= h_x(y) - U^{\sigma_x^G}(y) \geqslant 0 \quad \text{for } y \in G.$$

Since $U^{\sigma_x^G}(y)$ is harmonic in G we see that $g_y(x)$, regarded as a function of y, is an element of $\mathscr{G}_x$. Hence $g_y(x) \geqslant g_x(y)$. Similarly, $g_x(y) \geqslant g_y(x)$, and so (3) follows.

We have already given, in Theorem 2.41, the local form of the Riesz Decomposition. Armed with the Green function, we can give a more precise global form.

THEOREM 4.43. *Suppose that v is superharmonic and non-negative in a bounded open set G. Then*

$$v(x) = \int g_x^G(y)\ \mu(dy)+v^*(x)$$

where $\mu = C_p \Delta v$ and v^* is the greatest harmonic minorant of v in G.

Let M be an open set such that $\bar{M} \subset G$, and define f in $\partial(G \setminus M) = \partial G \cup \partial M$ by $f(\xi) = 0$ in $f(\xi)\ \partial G$ and in $= v(\xi)\ \partial M$.

Then the function w given by

$$w(x) = H_f^{G \setminus \bar{M}}(x),\ x \in G \setminus \bar{M}; \qquad w(x) = v(x),\ x \in \bar{M}$$

is quasi-superharmonic in G. This follows, since, first, w is superharmonic in $G \setminus \partial M$, next, $\liminf_{x \to \xi} w(x) = w(\xi)$ quasi-everywhere in ∂M and, since $v \in \mathcal{O}_f^{G \setminus \bar{M}}$ and $w \leqslant v$, then, for $\xi \in \partial M$,

$$w(\xi) = v(\xi) \geqslant \int v(\eta)\ \sigma_r^\xi(d\eta) \geqslant \int w(\eta)\ \sigma_r^\xi(d\eta).$$

Let $\hat{w}(x) = \liminf_{y \to x} w(y)$, so that $\hat{w}$ is superharmonic in G. Let $\nu = C_p \Delta \hat{w}$. Then (supp $\nu) \cap (G \setminus \bar{M}) = \phi$ and ν is the restriction of μ to $\bar{M}$. Writing $g_x^G = g_x$, consider

$$\int g_x(y)\ \nu(dy) = \int h_x(y)\ \nu(dy) - \int H_{h_x}^G(y)\ \nu(dy).$$

Since $g_x(y) = g_y(x)$ we have $H_{h_x}^G(y) = H_{h_y}^G(x)$. If we integrate the second integral on the right with respect to σ_r^x and invert the order of integration, we see that this second integral is harmonic in G.

Now, by Theorem 2.41, $\hat{w}(x) - \int h_x(y)\ \nu(dy)$ is harmonic in G and so $\hat{w}(x) - \int g_x(y)\ \nu(dy)$ is also harmonic in G. Furthermore, $\lim_{x \to \xi} \hat{w}(x) = 0$ q.e. in ∂G and, since $g_x(y) = g_y(x)$,

$$\lim_{x \to \xi} \int g_x(y)\ \nu(dy) = 0 \text{ q.e. in } \partial G$$

and so, by Theorem 4.38, and provided v is bounded,

$$\hat{w}(x) = \int g_x(y)\ \nu(dy) \text{ for } x \in G. \text{ Hence, } v(x) \geqslant \hat{w}(x) = \int_M g_x(y)\ \mu(dy),$$

and this holds for every M such that $\bar{M} \subset G$. Consequently

$$v(x) \geqslant \int g_x(y)\ \mu(dy). \tag{$*$}$$

If, now, v_n is not bounded let $v_n = \min\ [n, v]$ and let $\mu_n = C_p \Delta v_n$. Then we have $(*)$ with v_n and μ_n in place of v and μ. Denote this by $(*)_n$. Since μ_n is the restriction of μ to the subset of G in which $v(x) \leqslant n$ $(*)_n$ gives $(*)$ as n tends to ∞.

In proving this equality we have required only that v be superharmonic and non-negative. Now $v - v^*$ is superharmonic and non-negative, and $\Delta(v-v^*) = \Delta v$; so, for $x \in G$,

$$(v-v^*)(x) \geq \int g_x(y)\, \mu(dy).$$

Let $\{M_n\}$ be an increasing sequence of open sets with $\bar{M}_n \subset G$ and such that $G = \bigcup_{n=1}^{\infty} M_n$. By the Riesz Decomposition,

$$v(x) = \int_{M_n} g_x(y)\, \mu(dy) + u_n(x), \quad x \in M_n, \tag{4.14}$$

where u_n is harmonic in M_n.

For $x \in M_n$, $\{u_k\}_{k \geq n}$ is a decreasing sequence of non-negative functions harmonic in M_n, and so $u = \lim_{k\to\infty} u_k$ is harmonic in M_n for each n and hence harmonic in G. Since the integral in (4.14) converges increasingly to the integral over G as $n \to \infty$,

$$v(x) = \int_G g_x(y)\, \mu(dy) + u(x).$$

Hence $u \leq v$, and so $u \leq v^*$.

On the other hand

$$v(x) \geq \int g_x(y)\, \mu(dy) + v^*(x);$$

so $v^* \leq u$. Thus $u = v^*$ and we have the result.

If v is superharmonic and bounded below in G, then $v - \inf_G v$ satisfies the requirements of Theorem 4.43, and the greatest harmonic minorant of $v-k$ is v^*-k; so the result holds in this case also.

If v is superharmonic in G then v is superharmonic and bounded below in any M_n. It therefore has a greatest harmonic minorant v_n^* in M_n and

$$v(x) = \int g_x^{M_n}(y)\, \mu(dy) + v_n^*(x).$$

Now $\{v_n^*\}$ is decreasing, and so $v^* = \lim_{n\to\infty} v_n^*$ is either harmonic in G or identically $-\infty$ in some component of G. If the former holds, then

$$v(x) = \int g_x^G(y)\, \mu(dy) + v^*(x)$$

and v^* is the greatest harmonic minorant of v in G.

Since Axiom D holds, we may assert the equality of the best harmonic minorant and the greatest harmonic minorant for a locally bounded superharmonic function. In fact, more is true.

THEOREM 4.44. *The best and greatest harmonic minorant of any positive superharmonic function in a bounded open set G coincide.*

Let v^* denote the greatest and $\bar{v}$ the best harmonic minorant of v in G. Then $\overline{(v_1+v_2)} = \bar{v}_1+\bar{v}_2$ and $(v_1+v_2)^* = v_1^*+v_2^*$. Suppose for the moment, that v is continuous in a neighbourhood of ∂G. Then

$$\limsup_{x \in G, x\to\xi} v^*(x) \leqslant \limsup_{x \in G, x\to\xi} v(x) = v(\xi)$$

and so $v^* \in \mathcal{U}_v^G$ so that $v^* \leqslant H_v^G = \bar{v}$. But $v^* \geqslant \bar{v}$ and so, in this case, $v^* = \bar{v}$. By Theorem 2.41 we have

$$v(x) = \int_B h_x(y)\, \mu(dy)+u(x)$$

in any open ball B containing $\bar{G}$ where u is harmonic in B. Since u is continuous in a neighbourhood of ∂G, the equality of the minorants for v is equivalent to that for the potential only. If N is a neighbourhood of ∂G and

$$\mu = \nu_1+\nu_2+\nu_3$$

where ν_2, ν_3 are, respectively, the restrictions of μ to $N\setminus\partial G$ and ∂G then the potential of ν_1 is continuous in a neighbourhood of ∂G and so its minorants are equal; for sufficiently small N, $\nu_2(N)$ is arbitrarily small, and so, for fixed $x \in G$, the potential due to ν_2 is arbitrarily small and, consequently, both minorants are so. It follows that, for the potential of $\mu-\nu_3$, both minorants are equal.

We need therefore only consider the potential of a measure supported by ∂G. Suppose then, at the last, that $\operatorname{supp}\mu \subset \partial G$ and that

$$v(x) = \int h_x(y)\, \mu(dy).$$

Since v is harmonic in G,

$$v^*(x) = \int h_x(y)\, \mu(dy)$$

and

$$\bar{v}(x) = H_v^G(x) = \int v(z)\, \sigma_x^G(dz) = \int\int h_z(y)\, \mu(dy)\, \sigma_x^G(dz).$$

If we invert the order in the last double integral, we have

$$\begin{aligned} v^*(x)-\bar{v}(x) &= \int (h_x(y)-\int h_y(z)\, \sigma_x^G(dz))\, \mu(dy) \\ &= \int (h_y(x)-H_{h_y}^G(x))\, \mu(dy) = \int g_y(x)\, \mu(dy). \end{aligned}$$

Since $y \notin G$, $g_y(x) = 0$, and the result now follows.

§ 4.8. The Laplace Case for $\overline{\mathbb{R}}^p$

For $\overline{\mathbb{R}}^p$ Axiom 1 is obviously satisfied. So also is Axiom 2, since the open balls $B_r(a)$ are regular for the same reason as in $\mathbb{R}^p$, while $\overline{\mathbb{R}}^p\setminus B_r(a)$ is a regular set containing ω since the modified Poisson integral J_r^a provides the required solution to the Dirichlet Problem; then clearly the regular sets from a neighbourhood base for $\overline{\mathbb{R}}^p$. Axiom 3 is also satisfied, since, as we have pointed out, Theorem 2.24 also holds in $\overline{\mathbb{R}}^p$.

But we encounter a reverse when we come to consider Axiom P. By Corollary 2.3.1, any positive superharmonic function $\overline{\mathbb{R}}^p$ must be a constant. Since a constant is harmonic, it follows that there is no positive superharmonic non-harmonic function in $\overline{\mathbb{R}}^p$. In fact, then, $\overline{\mathbb{R}}^p$ provides an example of a space in which Axiom P does not hold; and indeed all the positive superharmonic functions are harmonic and proportional.

In fact, to deal with unbounded open sets, we consider $X = \overline{\mathbb{R}}^p \setminus \bar{B}_1(0)$. Clearly Axioms 1, 2, 3 hold in X. Also $v(x) = 1-|x|^{2-p}$ ($\log |x|$ for $p = 2$) is superharmonic, positive and non-harmonic in X, and so Axiom P is satisfied.

As to Axiom D, it is more convenient to observe that to obtain the key Theorem 4.38 we in fact require that the best and greatest harmonic minorants of a locally bounded potential for a relatively compact open set shall coincide. It will therefore be enough to show that this property holds in X. To do this, we introduce the Kelvin Transformation.

Let $\phi: \overline{\mathbb{R}}^p \to \overline{\mathbb{R}}^p$ be defined by

$$\phi(x) = x/|x|^2 \text{ when } x \neq \omega,\ x \neq 0;\quad \phi(\omega) = 0 \text{ and } \phi(0) = \omega.$$

Then ϕ is a one-one mapping and $\phi^{-1} = \phi$.

Given a function f positive in an open set $G \subset \overline{\mathbb{R}}^p \setminus \{0\}$, we define its Kelvin transform Kf in $\phi(G)$ by

$$(Kf)(y) = |y|^{2-p} f(y/|y|^2) \text{ when } y \neq 0;$$
$$(Kf)(0) = +\infty \text{ when } p \geqslant 3;\quad (Kf)(0) = f(\omega) \text{ when } p = 2.$$

Then, if $y \in \phi(G)\setminus\{0\}$, $(KKf)(y) = f(y)$; and f is continuous at y if, and only if, Kf is so at $\phi(y)$.

A function g harmonic near 0 has, by Theorem 2.42, the form

$$g(y) = k+\alpha h_0(y)+\sum_{n=1}^{\infty} Y_n(u)\,|y|^n+\sum_{n=1}^{\infty} Z_n(u)\,|y|^{2-p-n}$$

where $u = y/|y|$; and so

$$(Kg)(x) = kh_0(x)+\alpha+\sum_{n=1}^{\infty} Y_n(u)\,|x|^{2-p-n}+\sum_{n=1}^{\infty} Z_n(u)\,|x|^n$$

near ω, except that, for $p = 2$, k and α are interchanged. Now Kg is harmonic at ω if, and only if, $k = 0$ and $Z_n = 0$. Thus, for $p \geqslant 3$, functions harmonic at ω are transformed into functions having the form

$$\alpha h_0(y)+\text{harmonic function}$$

near 0, and functions of this latter form near 0 are transformed into functions harmonic at ω. For $p = 2$ harmonic goes into harmonic. Also, *if f is superharmonic and positive in an open set $G \subset \overline{\mathbb{R}}^p \setminus (0)$ then Kf is superharmonic and positive in $\phi(G)$.*

This may be proved by first supposing f twice continuously differentiable in $G \setminus (\omega)$. Then a straightforward, if tedious, exercise in elementary partial differentiation shows that

$$\sum_{r=1}^{p} \frac{\partial^2 f}{\partial x_r^2} \leqslant 0 \; (=0) \text{ if and only if } \sum_{r=1}^{p} \frac{\partial^2}{\partial y_r^2}(Kf) \leqslant 0 \; (=0).$$

After that, the fact that any superharmonic function is the limit of an increasing sequence of twice continuously differentiable superharmonic functions shows that Kf is superharmonic in $\phi(G) \setminus (0)$. Since Kf is l.s.c. at 0 and $(Kf)(0) \geqslant \mathscr{S}_r^0(Kf)$ we see that Kf is superharmonic in $\phi(G)$. Also, since f is not identically $+\infty$ in any component of G, Kf is not identically $+\infty$ in any component of $\phi(G)$.

The same argument will serve to show that, if f is harmonic in $G \setminus \{\omega\}$, then Kf is harmonic in $\phi(G) \setminus \{0\}$ (for $p = 2$, f being harmonic in G entails Kf harmonic in $\phi(G)$).

It also follows that if E is polar then $\phi(E)$ is polar, provided $0 \notin E$.

In what follows we shall concern ourselves with the case $p \geqslant 3$, since the case $p = 2$ can be dealt with by a simplification of the arguments presented. But, at the outset, we will remark on a difference between the case $p = 2$ and the case $p \geqslant 3$.

When $p = 2$, the set $\{\omega\}$ is polar since the function $\log |x|$ is positive and superharmonic in X and takes the value $+\infty$ at ω. *For $p \geqslant 3$ no set E containing ω is polar*. For even $\{\omega\}$ is not polar, since, if it were, we could, by Theorem 4.28(5), extend uniquely the function $|x|^{2-p}$ superharmonic in $\mathbb{R}^p$ to $\overline{\mathbb{R}}^p$. Since $\liminf_{x \to \omega} |x|^{2-p} = 0$ the extension would be $|x|^{2-p}$ in $\overline{\mathbb{R}}^p$. But this is not superharmonic in $\overline{\mathbb{R}}^p$.

Suppose now that p is a locally bounded potential in X and that $G \subset X$. Then $\phi(G)$ is a bounded open set in $\mathbb{R}^p$ and if $\omega \notin G$, then $0 \notin \phi(G)$. Let q^* be the greatest harmonic minorant of Kp in $\phi(G)$. Then Kq^* is harmonic in G and $\leqslant p$ and is maximal (for otherwise $KKq^* = q^*$ would not be maximal). Thus Kq^* is the greatest harmonic minorant p^* of p.

Suppose that $\bar{p}$ is the best harmonic minorant of p in G. Then, as we readily see,

$$K\bar{p} = \inf \{v \mid v \in \mathscr{H}^*_{\phi(G)} \text{ and } \liminf_{y \to \eta} v(y) \geqslant (Kp)(\eta) \text{ for } \eta \in \partial(\phi(G))\}$$

so that $K\bar{p}$ is the best harmonic minorant of Kp in $\phi(G)$. But, by Theorem 4.32, this is identical with q^*. Hence $\bar{p} = Kq^* = p^*$.

If $\omega \in G$, then $0 \in \phi(G)$ and $(Kp)(y) \sim |y|^{2-p} p(\omega)$ as $y \to 0$. Let q^* be the greatest harmonic minorant of Kp in $\phi(G) \setminus \{0\}$. Now

$(p(\omega))g_0^{\phi(G)} \leqslant (Kp)(y)$ in $\phi(G)$ and so $q^*(y) = |y|^{2-p}p(\omega)+u(y)$, where u is harmonic in $\phi(G)$ and maximal. Consequently, Kq^* will be harmonic in G, $\leqslant p$ and maximal, so that $Kq^* = p^*$. Let $\bar{p}$ be the best harmonic minorant for p in G. Then $\bar{p}$ is harmonic in G. Suppose first that $\bar{p}(\omega) > 0$. Then $(K\bar{p})(y) \sim |y|^{2-p}\,\bar{p}(\omega)$ as $y \to 0$, and so $\lim_{y\to 0} (K\bar{p})(y) = +\infty = (Kp)(0)$. Also $\lim_{y\to\eta} (K\bar{p})(y) = (Kp)(\eta)$, q.e. in $\phi(\partial G)$ since a similar relation holds between p and $\bar{p}$ in ∂G. Thus $K\bar{p}$ is the best harmonic minorant of Kp in $\phi(G)\setminus(0)$. It is therefore the greatest harmonic minorant and so $\bar{p}$ is the greatest harmonic minorant of p in $G\setminus(\omega)$ and so $\geqslant p^*$. Since $\bar{p} \leqslant p^*$, we now have $\bar{p} = p^*$.

If $\bar{p}(\omega) = 0$ then $\bar{p} = 0$ identically in G so $p(\xi) = 0$, q.e. in ∂G. Hence $(Kp)(\eta) = 0$, q.e. in $\partial(\phi(G))$ and so $K\bar{p} = 0$ identically. But then the greatest harmonic minorant of Kp is zero, and so $p^* = 0$. Thus $\bar{p} = p^*$.

It therefore follows that Theorem 4.38 applies in X. Given any open set $G \subset \mathbb{R}^p$ such that its complement has non-void interior, we may, by translation and change of scale suppose that $G \subset X$ and is relatively compact. And then we may solve the Dirichlet Problem for G.

On the other hand, if we are given an unbounded open set $G \subset \mathbb{R}^p$, the theory, in $\mathbb{R}^p$, does not furnish a solution for the Dirichlet Problem. But we can, in a sense, 'embed' the problem in $\overline{\mathbb{R}}^p$. Suppose $G \subset \mathbb{R}^p$ is such that $\mathbb{R}^p\setminus G$ contains a neighbourhood. Adjoin ω to $\mathbb{R}^p$ and consider G as a subset of $\overline{\mathbb{R}}^p$. Then either $\omega \in \partial G$ or $G \cup (\omega)$ is open.

In the first case the boundary data on ∂G (in $\mathbb{R}^p$) can be extended to boundary data on ∂G (in $\overline{\mathbb{R}}^p$) by setting, say, $f(\omega) = \limsup_{\eta\to\omega} f(\eta)$ and then Theorem 4.24 applies (we recall that, for $p \geqslant 3$, ω must be a regular point). If, in particular, $\lim_{\eta\to\omega} f(\eta)$ exists, there is a unique solution to the problem. Thus, for any $G \subset \mathbb{R}^p$ such that $\mathbb{R}^p\setminus G$ contains a neighbourhood, and for a continuous function f on ∂G such that $\lim_{\eta\to\omega} f(\eta)$ exists and is finite, we can solve uniquely the Dirichlet Problem.

In the second case ∂G is bounded, $\mathbb{R}^p\setminus\bar{G}$ is bounded and we are dealing with the exterior Dirichlet Problem for this latter set. We have seen that in the case of the open ball the problem is not well-posed if we specify only data on the boundary of the ball, unless we require the solution to be harmonic at ω also. We therefore have a choice; either we ask for a solution which is harmonic at ω also, which is to hand and unique, or we regard ω as a boundary point and specify the behaviour of f at ω. Thus we do have a unique solution to the exterior Dirichlet Problem posed in the form: f given on ∂G and $\lim_{|x|\to\infty} u(x) = k$, say.

Finally, for $p \geqslant 3$, we show that it is not necessary that the complement of G contain a neighbourhood. To do this we first need

LEMMA 4.45. *Suppose that* $G \subset \mathbb{R}^p$ *is open and that* ∂G *is not polar. Suppose that* $0 \in \partial G$ *and that* $u \in \mathscr{H}_G$ *is such that* $|u(x)| \leqslant M$ *for* $x \in G \setminus \bar{B}_\delta(0)$ *and*

$$\lim_{x \in G, x \to \xi} u(x) = 0, \text{ q.e. in } \partial G \setminus B_\delta(0).$$

Then, for $x \in G \setminus \bar{B}_\delta(0)$,

$$|u(x)| \leqslant M\delta^{p-2} |x|^{2-p}.$$

Let $H = G \setminus \bar{B}_\delta(0)$ and let w be associated with the set E of exceptional points in $\partial G \setminus B_\delta(0)$. Define g in ∂H by

$$g(\xi) \begin{cases} = \limsup_{x \in G, x \to \xi} u(x), & \xi \in \partial G \setminus \bar{B}_\delta(0) \\ = M, & \xi \in \partial B_\delta(0). \end{cases}$$

Consider the function

$$v_n(x) = M\delta^{p-2} |x|^{2-p} + n^{-1} w(x);$$

v_n is superharmonic in H and

$$\liminf_{x \in H, x \to \xi} (v_n(x) - u(x)) = \liminf_{x \in H, x \to \xi} v_n(x) - \limsup_{x \in H, x \to \xi} u(x)$$

which is $\geqslant 0$ for $\xi \in \partial H$.

Consequently, for $a \in H$, $u(a) \leqslant v_n(a)$. This holds for every n and, since we can have chosen w so that $w(a)$ is finite, we have

$$u(a) \leqslant M\delta^{p-2} |a|^{2-p}.$$

Arguing in similar fashion with $-u$, we now get the inequality in the lemma.

THEOREM 4.46. *Suppose that, for* $p \geqslant 3$, $G \subset \overline{\mathbb{R}}^p$ *is open and that* $\partial G \setminus \{\omega\}$ *is not polar in* $\mathbb{R}^p$. *Then, given* $f \in \mathscr{C}(\partial G)$, *there is a unique function* u *harmonic in* G *such that, except possibly in a set polar in* $\mathbb{R}^p$,

$$\lim_{x \in G, x \to \xi} u(x) = f(\xi), \quad \xi \in \partial G.$$

We suppose, as we may, by translation if necessary, that $0 \in \partial G$ and then suppose, in the first instance, that $\omega \notin G$. By the Tietze extension theorem (1.5) we may extend $\operatorname{rest}_{\partial G \cap \bar{B}_1(0)} f$ to $\bar{B}_1(0)$. Call this extension f^*. Then $\sup |f^*| \leqslant \sup_{\partial G} |f|$.

Let $B_n = B_{1/n^2}(0)$, and let $H_n = G \setminus \bar{B}_n$. Then, since the complement of H_n contains a neighbourhood, we can find a function u_n harmonic in H_n such that

$$\lim_{x \in H_n, x \to \xi} u_n(x) \begin{cases} = f(\xi) \text{ q.e. in } \partial G \setminus \bar{B}_n \\ = f^*(\xi) \text{ q.e. in } \partial B_n. \end{cases}$$

Consider $v_n = u_{n+1} - u_n$. It is harmonic in H_n and

$$\lim_{x \in H_n, x \to \xi} v_n(x) = 0, \text{ q.e. in } \partial G \setminus \bar{B}_n$$

and

$$| v_n(x) | < 2 \sup_{\partial G} | f |, \quad x \in H_n.$$

Hence, by Lemma 4.45, for $x \in H_n$,

$$| v_n(x) | \leqslant 2 \sup_{\partial G} | f | \, n^{4-2p} \, | x |^{2-p}.$$

In consequence $\{u_n(x)\}$ is uniformly convergent in $G \setminus B_\rho(0)$ for any fixed $\rho > 0$, and so $u(x) = \lim_{n \to \infty} u_n(x)$ is harmonic there. Furthermore, since $\lim_{x \in G, x \to \xi} u_n(x) = f(\xi)$, q.e. in $\partial G \setminus B_\rho(0)$ the same holds for $u(x)$. Since this holds for every $\rho > 0$ we have, finally,

$$\lim_{x \in G, x \to \xi} u(x) = f(\xi), \text{ q.e. in } \partial G.$$

Now suppose $\omega \in G$. Then ∂G is bounded. By what we have just proved we can, given k, find a function u harmonic in $G \setminus \omega$ such that

$$\lim_{x \in G, x \to \xi} u(x) = f(\xi) \text{ q.e. in } \partial G; \qquad \lim_{x \to \omega} u(x) = k.$$

Then u is harmonic in a neighbourhood of ω and so

$$u(x) = k + \alpha \, | x |^{2-p} + \sum_{n=1}^{\infty} Z_n(u) \, | x |^{2-p-n}$$

there. Now α (and z_n) depend on k. For large positive k, α is negative (otherwise u would attain a maximum in G), and for large negative k, α is positive (otherwise u would attain a minimum in G). Also, for sufficiently large

$$\mathscr{S}_r^0(u) = k + \alpha \, | r |^{2-p}.$$

Since u depends continuously on k, so does $\mathscr{S}_r^0(u)$, and so α depends continuously on k. Also, fixing r, we see that α decreases with increasing k. Consequently, we can find one, and only one, value of k, k_0 say, for which $\alpha = 0$. If we choose this value of k, the resulting u is harmonic at ω also, and u is then the required solution.

As to uniqueness, when $\omega \notin G$, we can argue as in the uniqueness proof of Theorem 4.38 since, as ω is regular, we shall be seeking a function superharmonic in $\mathbb{R}^p$ (and not in $\overline{\mathbb{R}}^p$) associated with the set of irregular points, and this is to hand. Then, if $\omega \in G$, the method by which we pass from the case $\omega \notin G$ to the case $\omega \in G$ shows that here, too, u will be unique.

The result is also true for $p = 2$, but the method of proof fails. The reader is referred to Brelot [4] for a discussion which does not make $p = 2$ an excluded case.

REFERENCES

BAUER, H. 1962. Axiomatische Behandlung des Dirichletschen Problem für elliptische und parabolische Differentialgleichungen. *Math. Annln*, **146**, 1–59.

BRELOT, M. [1] 1952. La Théorie Moderne du Potentiel. *Annls Inst. Fourier Univ. Grenoble*, **4**, 113–140.

[2] 1959. *Éléments de la Théorie Classique du Potentiel.* "Les Cours de Sorbonne" Centre de Documentation Universitaire, Paris.

[3] 1960. *Lectures on Potential Theory.* Tata Institute of Fundamental Research.

[4] 1944. Sur le rôle du point a l'infini dans la théorie des fonctions harmoniques. *Annls scient. Éc. norm. sup., Paris*, **61**, 301–332.

[5] 1940. Points irreguliers et transformations continues en théorie du potentiel. *J. Math. pures appl.*, **19**, 319–337.

CARTAN, H. 1946. Théorie generale du balayage en potentiel newtonien. *Ann. Univ. Grenoble*, **22**, 221–280.

CHOQUET, G. 1952–4. Theory of Capacities. *Annls Inst. Fourier Univ. Grenoble*, **4–5**, 121–130.

DU PLESSIS, N. [1] 1964. Two counter-examples associated with the Dirichlet Problem. *Q. Jl Math.*, **15**, 121–130.

[2] 1966. Some further examples associated with the Dirichlet Problem. *Q. Jl Math.*, **17**, 1–6.

FROSTMAN, O. 1935. Potentiel d'Équilibre et Capacité des Ensembles avec quelques applications à la théorie des fonctions. *Meddn. Lunds Univ. mat. Semin.*, **3**, 1–118.

FUGLEDE, B. 1960. On the theory of potentials in locally compact spaces. *Acta math., Stockh.*, **103**, 140–215.

HERVÉ, R. M. 1962. Recherches axiomatiques sur la Théorie des Fonctions surharmoniques et du Potentiels. *Annls Inst. Fourier Univ. Grenoble*, **12**, 415–571.

KELLOGG, O. D. 1929. *Foundations of Potential Theory.* Springer.

LANDKOV, N. S. 1966. Основы Современной Теояии Потенциала. "Наука". Moscow (1966).

LEBESGUE, H. 1913. Sur des cas d'impossibilité du problème de Dirichlet. *C. r. Séanc. Soc. math. Fr.*, **17**.

RUDIN, W. 1966. *Real and Complex Analysis.* McGraw Hill.

SAKS, S. 1933. *Theory of the Integral. Monografie mat.*, Warsaw.

ZAANEN, A. C. 1958. *An Introduction to the Theory of Integration.* North Holland.

ZAREMBA, S. 1911. Sur le principe de Dirichlet. *Acta math., Stockh.*, **34**, 293–316.

INDEX